The Sun and How to Observe It

P9-BJK-706

For further volumes:
www.springer.com/series/5338

The Sun and How to Observe It

 Springer

Jamey L. Jenkins
Homer, IL, USA
jenkinsjI@yahoo.com

ISBN: 978-0-387-09497-7   e-ISBN: 978-0-387-09498-4
DOI 10.1007/978-0-387-09498-4

Library of Congress Control Number: 2008939373

Printed on acid-free paper

springer.com

Acknowledgements

I extend my sincere appreciation to Dr. David Hathaway, David Williams, and Fred Espenak of NASA for permission to use data appearing on their web pages and also to Mats Löfdahl of the Royal Swedish Academy of Sciences. Special thanks is given to Richard Hill of the Lunar and Planetary Laboratory, University of Arizona, for his support and to Gordon Garcia, Art Whipple, Eric Roel, Howard Eskildsen, Steve Rismiller, Greg Piepol, Vincent Chan, Christian Viladrich, Gema Araujo, and Lex Lane for use of their spectacular imagery. Brad Timerson of the Association of Lunar and Planetary Observers Solar Section computed and supplied the solar Ephemeredes found in the back of the book.

My gratitude goes out to Amber Eldridge for information regarding skin protection and the Sun, and also to Ray Cash and Robert Hess, who provided outlines of their dedicated solar telescopes. And to all my correspondents on the Internet who share this wonderful hobby with the world and me.

Mostly, though, I offer my deepest appreciation to my wife, Mary, and my immediate family for permitting me the time out of our busy lives to attempt to put on paper an encouraging word or two that inspires what should be a thrilling experience for anyone possessing the desire to observe the Sun. Observing most astronomical objects and particularly the Sun requires unending patience, continual development of your astronomical eye, and the quenching of a relentless thirst for knowledge. The Sun with its multi-faceted face tests and satisfies these essentials.

Clear Skies
Jamey Jenkins IL, USA

Contents

About the Author

Jamey Jenkins has been a regular contributor to the Sunspot Program of the American Association of Variable Star Observers (AAVSO) since 1990 and an active observer for the Association of Lunar and Planetary Observers (ALPO) Solar Section since 1998. He has been Assistant Section Coordinator of that group for the last 3 years. An astronomy enthusiast since youth, he also has published numerous articles in *The Strolling Astronomer* and images in *Sky & Telescope*.

Living among the corn and soybean fields of mid-America affords wide-open views of the night and daytime sky. When not exploring the Moon, planets, or his favorite celestial body, the Sun, he earns a living as a digital pre-press specialist with R. R. Donnelley, the largest commercial printer in North America at its Crawfordsville, Indiana, facility.

Jamey and wife Mary are the parents of four adult daughters and four grand-children. "We are a family of varied interests and talents. We are printers, health care professionals, musicians, teachers, historians, social service providers, and computer technology specialists. Astronomy, though, has touched only two of us. My son-in-law Chris and I are followers of Galileo and Copernicus, spending our spare time enjoying the beauty of the heavens and sharing it with anyone that is curious enough to peek through the telescope," says Jamey.

Introduction

Over four decades ago an amateur astronomer browsing the stacks at his or her local library might have come across a copy of William Baxter's book, *The Sun and the Amateur Astronomer*. This English author's text gave aspiring solar astronomy buffs a look into the how and why of techniques used for observing the Sun. Baxter carefully painted a picture of how an amateur astronomer, using only a modest telescope, sketch pad, and sheet film camera, could leisurely record solar activity. Over the years a number of devotees, including myself, found Baxter's work invaluable in the pursuit of their solar astronomy hobby, his being the first book of its kind written for amateur studies of the Sun.

An amateur observer from that era would hardly be able to imagine the current astronomy scene. Observational astronomy has experienced a complete revolution! For the most part pencil and paper, vital in Baxter's time, are now relegated to note taking. The electronic sensor has replaced film, and advanced video techniques offer the most promise for those attempting to record the finest solar detail in their photographs.

Another surprise for the earlier observer might be the availability of commercial telescopes dedicated specifically to solar observations. In the past monochromatic observing, done by utilizing a thin slice of light from the solar spectrum, was available only to the craftsman capable of building the complex, delicate instruments needed to perform such observations. These instruments, the spectrohelioscope and monochromator, were expensive and often beyond the skill of the typical telescope maker to construct. Since that time, the availability of solar telescopes and filters for Hydrogen-alpha and Calcium-K observing have awakened an interest in daytime astronomy to a whole new generation of observers. Using an off-the-shelf solar telescope, today's amateur astronomers coupled with an inexpensive computer webcam are producing time-lapse movies of chromospheric activity that was previously only the domain of a professional astronomer located at a high-altitude solar observatory. Never before have such opportunities existed for amateur observers. This is truly an exciting time to be a solar astronomy hobbyist.

With this book we hope to project the sense of excitement that so many observers experience when we point our telescopes sunward. If you are new to solar astronomy, you should become educated on how to safely explore the Sun. Veteran observers could find in these pages a new twist to an old technique that allows seeing the Sun in a different way.

As a variation on the hobby of star gazing, solar observing provides an alternative to late nights, cold fingers, and fumbling in the dark trying to locate that expensive eyepiece you've just dropped in the dew-soaked grass. All events happening on the Sun are unique and never will be repeated exactly. This is much of what attracts individuals to solar astronomy and is the reason there is a scientific value to each of your observations. Whether you follow the growth and decay of a sunspot group, the rapid emergence of a solar flare, or the spray of an erupting prominence at the Sun's limb, one fact is certain: the Sun will always present a uniquely different face, each and every day.

In order to appreciate the Sun and its ever-changing face, it's valuable to have an understanding of what it is, how it works, and how it relates to our world. The Sun *is* a star, a sphere of glowing hot gases, one star in a massive collection called the Milky Way galaxy. Enormous pressures exist inside the Sun, creating an environment unlike anything we could possibly experience on Earth. Nuclear forces that influence conditions on our Earth and the other planets in the Solar System are released deep within the Sun's core. The first part of this book will give an overview of these topics. We will begin by looking at the differences and similarities between the Sun and other stars, how the Sun was born, and how energy makes its way from the Sun's core to our backyards. Once that basic foundation is established, the discussion will shift to how an amateur astronomer of the twenty-first century observes the Sun. Together, we will explore the cavalcade of features to be seen in white and monochromatic light and the instruments that can be used to safely observe them. In the latter part of the book, we will review modern techniques for rendering and sharing your solar observations with the world, itself a hobby within a hobby.

A word of CAUTION to prevent the uninitiated from rushing out into the daylight and directing their telescopes skyward. Solar observing can be a very dangerous activity unless certain safety guidelines are followed, a theme you will find repeated throughout this book. The Sun emits huge quantities of heat, light, and radiation, which the solar observer must respect at all times. The atmosphere and magnetic field of Earth fortunately act as a shield for much of the radiation; the daily danger to the Earth-bound astronomer is in the brightness and the infrared and ultraviolet light of the Sun. These invisible wavelengths must be filtered out, and the intensity of the illumination reduced to an acceptable level for safe visual studies to be conducted. **Without these necessary precautions, blindness of the observer will result.** Of course, this topic will be discussed in greater detail in the following chapters. Regardless, the author and publisher cannot be held responsible for the careless actions of any solar observer disregarding safety procedures. The rule of thumb regarding solar observing is this: Always err on the side of safety when observing the Sun. Do that, and you'll be able to enjoy many years of watching one of nature's most magnificent spectacles from your own backyard.

Chapter 1

The Sun, Yesterday and Today

One of Millions

As a young man I often took a nightly stroll down a pathway that led to a meadow far from my home. Looking skyward on many of those dark summer nights I studied a heaven full of silent, twinkling stars that on occasion reminded me of a smattering of jewels flung across a dark velvet cloth. I would see a pale diffuse web of light rising in the northeast near Cassiopeia that stretched clear to the southern horizon. Exploring this diffuse web with a small pair of binoculars revealed to me that it was composed of a countless number of individual stars. But to the naked eye this pale light was deemed to be the arm of a spiral galaxy, snaking its way to a hub located in Sagittarius. In fact, every naked-eye star I could see from that country pathway was part of this galaxy's family. The evening sky seemed to be saying, "Welcome to the Milky Way."

The Milky Way is the galaxy where we live, orbiting on a sphere shaped platform we call Earth, about a typical star that long ago our ancestors chose to call Sol, or the Sun. There was a time, only several hundred years ago, when people thought Earth was the center of the universe and that all celestial bodies were revolving around it. And why not? Isn't that how it appears to the untrained eye? Today, we know the truth. Earth is one of many thousands, if not millions of bodies, large and small that circles the Sun. Furthermore, this assemblage of gas, liquid, dust, ice, and rock we call our Solar System orbits the galactic nucleus, the hub of the Milky Way.

Careful observations by astronomers tell us that the Sun is located about one-third of the way from the outer edge of the Milky Way, a distance of about 25,000 light-years from the center of the galaxy. One light-year measures 9.46×10^{12} km. Our galaxy has an overall diameter in the vicinity of 80,000 light-years. Since the Solar System travels at nearly 230 km/s through space, it takes the Sun close to 200 million years to complete a circuit of the galaxy. Scientists say that the Milky Way contains hundreds of millions of other stars besides the Sun, some similar but many different (Figure 1.1). Our view of an Earth-centered universe has changed dramatically in the last 500 years!

What Exactly is the Sun?

The Sun is a typical star, a giant sphere-shaped ball of gas that through nuclear reactions releases energy in its core. Due to the great distances found between stars, most appear similar when viewed through a telescope. In reality, though,

J.L. Jenkins, *The Sun and How to Observe It*, DOI 10.1007/978-0-387-09498-4_1,
© Springer Science+Business Media, LLC 2009

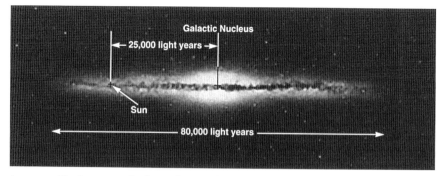

Figure 1.1. The Sun is one of millions of stars within the Milky Way Galaxy. In this artist's rendition, the Sun is situated about 25,000 light years from the center of the galaxy.

stars have a wide range of individual characteristics. All stars including the Sun vary from one another in color, temperature, and brightness, not to mention mass, composition, and age.

Although the Sun is 1.3 million kilometers in diameter and provides a reasonable angular size against our sky, on average 32 arc minutes, stars are almost always seen as virtual points of light. This is true regardless of the fact that some stars are absolute giants when compared to our Sun. Take the star Antares in the constellation Scorpius. Antares is a red supergiant about 520 light-years distant from our Solar System and about 230 times larger in diameter than the Sun. In spite of its gigantic size, Antares is still seen as a point through our telescopes.

The distance separating the Sun from Earth is, on average, 150 million kilometers. Remember, we are actually closer to the Sun during the month of December and farther away in June, an indication that Earth's orbit is not circular but elliptical. Compare the Sun's distance from us to the next nearest known star Proxima Centauri, at a distance of 4.2 light-years or to Sirius, the brightest star in the night sky, at 8.6 light-years. One analogy that puts such astronomical distances as these in perspective is this: "Visualize the Earth-Sun system with Earth represented by the tiniest of pebbles and the Sun as a large marble, separated by a distance of only 1 m. Now, with this scenario, the nearest star Proxima Centauri would be over 265 km away!" As you can see, the Sun is unique because of our nearness to it and its great distance from the other stars.

Another significant difference between the Sun and stars is brightness. The system called magnitude defines the brightness of a celestial object. On the magnitude scale, objects assigned larger numbers are fainter; those with smaller numbers appear brighter. Each step of magnitude is designed to be about 2.512 times brighter or fainter than the preceding step. In other words, stars of 2nd magnitude look about one hundred times brighter than stars of 7th magnitude ($1 \times 2.512 \times 2.512 \times 2.512 \times 2.512 \times 2.512$).

There are two basic kinds of magnitude, absolute and apparent. Absolute magnitude is a measurement of the light received from an object when placed at a set distance of 32.6 light-years from Earth. Absolute magnitude describes a body's true brightness. Apparent magnitude measures the amount of light we see regardless of an object's distance from Earth. This is ordinarily the magnitude assigned to an object; it measures how bright a celestial body appears to us. The brightest object in our sky, with an apparent magnitude of −26.8, is the Sun. The Moon at full has a magnitude of −12.7, Sirius shines at −1.4, and Polaris, the North Star, is magnitude +2.1.

Generally, stars of about 6th magnitude are as faint as can be glimpsed by an observer without optical aid. The dimmest objects found during the Hubble telescope's Ultra Deep Field Survey are on the order of 31st magnitude; some stars can appear very faint indeed.

Because a star's color is directly related to its surface temperature, and stars vary from relatively cool to very hot, a virtual rainbow of starlight is visible in the night sky. This relationship between temperature and color, called, Wien's Law, states that in principle the dominant emission wavelength of a blackbody (wavelength of its color) multiplied by its temperature must equal a specific numerical factor. Black-bodies are objects that reflect no light, but do absorb and re-emit radiation. Stars are blackbodies. The peak output in a star's spectrum determines its dominant emission wavelength; it is this wavelength or position in its spectrum that produces the star's color, and serves as a factor, according to Wien's Law, in its surface temperature. This makes sense if you consider a fireplace poker left in a fire. The poker becomes a glowing red color (called red-hot), because of the specific temperature it has attained. Stars with a specific temperature glow a particular color.

Table 1.1. Pertinent facts regarding the physical makeup of the Sun. Data courtesy of NASA

Diameter: 1,391,980 km (109 Earth diameters)
Mass: $1,989,100\times10^{24}$ kg (333,000 Earths)
Volume: $1,412,000\times10^{12}$ km^3 (1,304,000 Earths)
Visual magnitude: −26.74
Absolute magnitude: +4.83
Spectral type: G2V

Distance from Earth:	Minimum	147,100,000 km
	Mean	149,600,000 km
	Maximum	152,100,000 km
Apparent diameter:	Minimum	31.4 min of arc
	At 1 A.U.	31.9 min of arc
	Maximum	32.5 min of arc

Central pressure: 2.477×10^{11} bar
Central temperature: 1.571×10^7 K
Central density: 1.622×10^5 kg/m^3
Central composition: 35% H, 63% He, 2% C, N, O...

Photosphere pressure (top): 0.868 mb
Photosphere temperature (top): 4400 K
Photosphere effective temperature: 5778 K
Photosphere temperature (bottom): 6600 K
Photosphere composition: 70% H, 28% He, 2% C, N, O...

Sidereal rotation period: 25.38 days
Synodic rotation period: 27.27 days

Age: 4.57×10^9 years

The color of the Sun is yellow-white, similar to that of the star Altair, in the constellation of Aquila. Contrast that with the blue of Bellatrix in Orion the Hunter or the orange of Aldebaran in Taurus the Bull. Stars with a blue hue, such as Bellatrix, have surface temperatures of 20,000–35,000 K (water boils at about 373 K). Aldebaran's cooler surface temperature is approximately 4000 K. Our Sun checks in at about 5800 K.

The Sun, without a doubt, is the king of our Solar System. We are dependent on its existence to provide warmth, light, and inevitably life to our world. Understand also that the Sun has its place among the stars of the heavens. It is a typical star, and because of our ideal location, we have a front row seat to witness phenomena that would evade us on all other stars in the universe. This vantage point near the Sun helps us to contemplate, and understand, the differences between the Sun and the other stars (Table 1.1).

The Origin of the Sun

The Sun and its Solar System, we believe, began as a vast cloud of gas and dust called the solar nebula. It is speculated that the nebula had a mass of two or three times that of the Sun and a diameter of at least one hundred times the Earth to Sun distance. This cloud was composed of a number of elements including hydrogen, helium, carbon, nitrogen, oxygen, neon, magnesium, silicon, sulfur, and iron. Present but not in abundance were nickel, calcium, argon, aluminum, and sodium. Several other elements in only trace amounts were also to be found, including gold.

Since the beginning, which we call the Big Bang, hydrogen and helium always have been the most prolific elements of the universe, totaling almost 98% of its combined mass. The other elements of the solar nebula were produced inside the first stars, through nuclear processes or by experiencing the destructively powerful ending to a star, called a supernova.

Dust particles, found within the solar nebula, were likely coated with an "ice" created by some elements that were condensing in the frigid temperatures of that time. Tugging on these dusty ice particles, gravity would cause a general tendency of their movement toward the center of the solar nebula. In time, as these particles collected, gravity-induced density and pressure would increase in the central region of the solar nebula. Inside this so-called protosun it became crowded; little room existed between atoms, causing them to repel one another, producing thermal energy, or heat. This process, the turning of gravity's energy to heat, is called the Helmholtz contraction. To prevent all matter from being drawn into the center by gravity, with no planets forming, angular momentum or a rotation of the solar nebula had to somehow be present. Rotation of the nebula may have been a natural characteristic, or it could have been the result of a passing shockwave from a nearby supernova explosion.

Eventually, the pressure and temperature resulting from the contracting gas and particles inside the solar nebula would have reached a point where the new protosun could "switch on" and begin to glow.

Although gravity was responsible for initiating the early Sun, the process of Helmholtz contraction was not one that could sustain a star's appetite for energy. A different process had to be at work, fueling the Sun as we see it today. What was it? The answer to that question came in 1905 from Einstein's theory of relativity, which states that energy and mass are interchangeable. The equation, $E = mc^2$, tells us that energy (E) is equivalent to an object's mass (m) times the speed of light (c) squared. What does this mean to us in our understanding of how the Sun is powered? Simply said, a tiny amount of matter can be converted to a stupendous amount of energy! Physicists after a time proposed that given the right conditions of temperature and pressure, as are found in the Sun, hydrogen atoms could fuse together, forming helium, with a portion of the Sun's mass being released as energy, fueling the solar furnace. This is just what the Sun does.

How the Sun Works

All stars, including the Sun, are energized by nuclear reactions deep within their core. The pressure and temperature in the Sun's core is so extraordinary high that four hydrogen protons fuse to become a single helium nucleus. It's estimated that the pressure in the core is nearly 340 billion times the air pressure at sea level on Earth. Because of such intense pressure, temperature in the core of the Sun exceeds 15 million Kelvins.

Gas within the core has a density many times that of lead, and conditions are so extreme that electrons are stripped from their atoms. This process of separation of electrons from an atom is called ionization; an atom with one or more electrons missing is called an ion. Atoms within the core of the Sun are totally ionized, and in this state a gas is called a plasma, which is a brew of ions and electrons that react energetically with magnetic fields. Stars having a mass equal to or less than the Sun go through a process of converting hydrogen to helium; this process is referred to as the proton-proton cycle. Stars with a greater mass than the Sun also convert hydrogen to helium, but through a different process called the CNO cycle.

The proton-proton cycle results in the regeneration of millions of tons of hydrogen to helium in the Sun every second. As time goes by, the Sun will become lighter and exhaust the hydrogen it has been burning for billions of years. One day, after the hydrogen is gone, the outer layers of the Sun will be blown away to form what is ironically called a planetary nebula. Sadly, our planet will cease to exist, and in the end, the Sun will remain an insignificant white dwarf star. No need for immediate concern, however; there is an estimated 5 billion additional years of hydrogen available. The hydrogen to helium fusion process has already been underway for nearly 4.6 billion years (Figure 1.2).

Energy from the Inside Out

Thermonuclear fusion within the Sun's core is the source of solar power. Tremendous amounts of energy are being released, yet the Sun does not explode like an atomic bomb. Because of the forces of equilibrium, the Sun remains in a relatively steady state. Outward pressure from the compression of gas prevents gravity from causing the outer layers of the Sun to collapse into the core. This balancing act of pressures is called hydrostatic equilibrium. Likewise, the nonstop conversion of hydrogen to helium is happening at a uniform rate. There is no sputtering, starting, or stopping of the hydrogen burning. This continuous fuel-in, energy-out rate is called thermal equilibrium. Without these two balancing principles, the Sun as we know it couldn't exist.

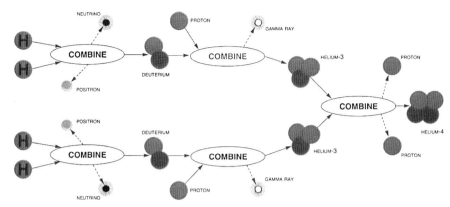

Figure 1.2. Energy is created within the core of the Sun through a process called the proton-proton cycle. In this chain-reaction, two hydrogen protons combine to form deuterium, an isotope of hydrogen. During the first phase of the conversion process, one of the protons becomes a neutron, resulting in its casting off a neutrino and a positron. Next, the deuterium nucleus combines with another proton. This second reaction results in energy being released as a gamma-ray photon. The new helium$_3$ nucleus then combines with another helium$_3$ nucleus to produce a helium$_4$ nucleus, and two protons are released as a result.

But how does the energy get from the solar furnace in the core to the region we call the photosphere and beyond? To begin, we must recognize that the Sun is a body consisting of a number of zones or layers. Imagine for a moment the cross-section of a baseball. At the center of a baseball is a smaller hard rubber core, which is wound with a string like material, woven tightly to build up the height of the ball to its proper circumference. Around this again is stitched the outside covering of the ball. A baseball is constructed in layers, and so is the Sun. From the solar interior to the exterior, we find the core, then the radiative and the convection zones. Immediately above the convection zone we find the photosphere, the first layer of the solar atmosphere. The light we see from the Sun emanates from the photosphere.

Radiative Zone

Progressing outward, directly above the core and extending to a point that is nearly 70% of the way from the center to the solar radius is the radiative zone. Near the bottom of this layer the temperature is about 8,000,000 K, and the density several times that of lead. Gamma ray photons are a form of energy released in the Sun's core during nuclear fusion. Photons are light rays. As the photons flow from the core into the radiative zone, the gases present there will absorb and re-radiate the rays. The general tendency of photons is to depart from the hot interior toward the cooler photosphere. Inside the Sun, however, it is very crowded. The high-energy gamma ray photons are knocked from side to side, absorbed, re-emitted, and sometimes take a path back toward the center, spending a hundred thousand or more years finding their way through this zone.

Convection Zone

Above the radiative zone and below the photosphere, having a depth of about 210,000 km, is a layer termed the convection zone. Here, energy is transported through the passage of plasma from deep in the zone to the upper layer, the photosphere. As the hot gases rise, they cool and fall back toward the Sun's interior in a process known as convection. The analogy of a bubbling pot of oatmeal often comes to mind when describing solar convection. Heat, generated at the bottom of a pot, collects in pockets within the oatmeal. A heated pocket then begins to rise to the surface of the oatmeal, transferring energy and producing a "bubbling" in the pot. On the photosphere a similar effect can be seen. The photons produced in the core and passed through the radiative zone create convection cells, causing them to rise to the solar surface. On the Sun, each 1000-km-diameter convection cell, called a granule, makes its way to the top of the convection zone at nearly 1500 km/h. Releasing energy in the photosphere, the granule cools as the gas flows back to the solar interior along the granule's outer wall. These darker, cool outer walls give granules their unique kernel shape.

Granules cover the entire visible surface of the Sun, totaling several million at any given time. The life of a granule is "brilliant" but short, each lasting perhaps 5–10 min, only to be replaced by the next bubble emerging from deep in the convection zone (Figure 1.3).

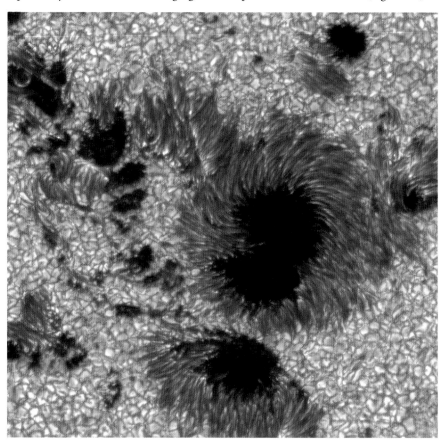

Figure 1.3. High-resolution image of a sunspot group with granules surrounding it. Obtained with the 1-m solar telescope at La Palma, Spain. Courtesy of Royal Swedish Academy of Sciences.

There is another movement of plasma in the convection zone that has been shown to occur from the equatorial region of the Sun to the polar areas. This movement is called the meridional flow and may be responsible for the migration of sunspot groups toward the equator as a solar cycle progresses. The solar cycle is roughly an 11-year-long cyclic period of activity on the Sun. As time advances, indicators of an active Sun, such as sunspots and flares, increase in quantity, until a peak occurs and then activity begins to decline. Gradual movement of plasma due to the meridional flow creates a circuit or loop, traveling from the solar equator to a point near the Sun's poles. In the polar regions, the plasma makes a curl under toward the bottom of the convection zone before resuming a slower return trip to the equator.

Sunspot migration from the higher solar latitudes to the equator is tied to the belief that groups are magnetically anchored to the lower region of the convection zone. The slow pace of the meridional flow is conjectured to be a factor for sunspots surfacing closer to the equatorial regions as a solar cycle advances.

Photosphere

The photosphere (sphere of light) is the beginning of the solar atmosphere and the lowest level we can see visually into the Sun. Below this layer, gas is so opaque that it is impossible to see through. However, in the photosphere watchful eyes see granulation, sunspots, and near the solar limbs, wispy material called faculae. Activity in the photosphere follows the 11-year ebb and flow of the solar cycle.

The photosphere is akin to the covering on the baseball referred to earlier. When the surface of the Sun is spoken of, this is the layer meant. Of course, the Sun really has no "surface," being gas, but because this region is the emitter of most of the light we see, it appears to be the surface. Approximately 500-kilometers thick, the temperature at the lower boundary is about 6600 K, while at the top it has dropped to nearly 4400 K, with a pressure less than 1 mb.

Photons from the inner zones reaching the photosphere are set free and shoot into space. Streaming out of the Sun, they make it a brilliant and dangerous object to watch without sufficient eye protection. It's marvelous to contemplate that the light we see leaving the Sun today started its course from the core and through the several outer layers many thousands of years ago (Figure 1.4).

Magnetic Fields

Below the Sun's surface, atomic forces from the pressure of gas prevail, but in the photosphere and beyond magnetism takes over as the dominant force. The Sun's magnetic field is the result of rotation and convective motions within the solar interior. Helioseismologists, astronomers who study low-frequency sound waves originating in the Sun, tell us that the radiative zone and core rotate like that of a solid body, with a period of about 27 days. Rotating differentially, the convection zone and upper layers experience a rotation rate near the equator of about 25 days; near the polar regions, acting as if the it were made of liquid, the Sun's rotation rate is about 36 days.

Last century, astronomer Horace Babcock created a theory that helped explain the appearance of sunspots within the photosphere. According to Babcock, the

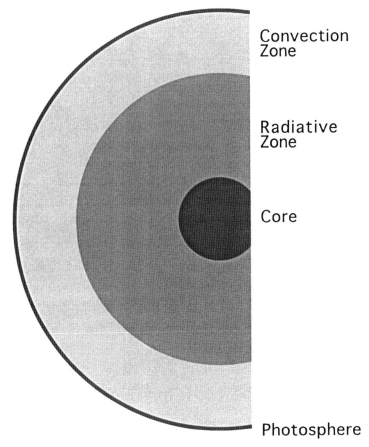

Figure 1.4. Internal layers of the Sun.

magnetic field of the Sun is influenced by plasma flow inside the solar interior. Shearing between the solid acting radiative zone and the fluid-like convection zone also contributes to the magnetic field. This region between the two zones is called the tachocline.

The nature of a plasma as it moves is to create a magnetic field. The lines of the Sun's magnetic field run parallel to the axis of rotation, from pole to pole in a north-south direction. Differential rotation in the convection zone wraps the magnetic field round and round the Sun, similar to how the baseball model is wrapped with string about its core. This stretching or wrapping occurs because the magnetic lines are being dragged along with the charged particles of plasma.

Convection is also at work transferring energy from the radiative zone to the photosphere, with a vertical boiling motion. This vertical movement of plasma causes a tangling of the field lines. The tangled field lines create an increase in strength while developing kinks in their paths. A strand or kink of magnetic field suspended in the convection zone is called a flux tube. Smaller flux tubes pop through the solar surface at bright points known as filigree, which have a diameter around 150 km. Much larger flux tubes are dark and give birth to pores and sunspots.

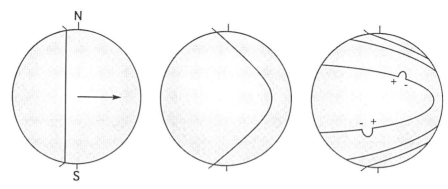

Figure 1.5. Sunspot theory says that a magnetic field line runs from north to south on the Sun, as at the left. The differential rotation of the Sun illustrated in the center causes the magnetic field to stretch and wrap around the Sun. In the right diagram, after many rotations the magnetic field has become tangled with other field lines; strengthening, it rises to the photosphere to form an active region.

When a tube has reached strength sufficient to cause it to rise to the surface and burst through, leaving a magnetic impression on the photosphere, the tube is called an active region. Like the horseshoe magnet has a north and south pole, so does an active region. The magnetic field projects above the photosphere and arches from one polarity (+) to the other (−). The leading and following magnetic polarities of all active regions are identical, depending on which hemisphere contains the region; the polarities are reversed in the opposite hemisphere. In a process not fully understood, these fields switch polarities in step with the 11-year cycle of sunspot activity. Two complete solar cycles, therefore, make one magnetic cycle, a period of about 22 years.

Sunspots, with their dark umbral and lighter penumbral regions, appear within magnetic fields on the photosphere because convection is stifled. Only smaller amounts of energy via convection are able to reach an active region, which becomes cooler than its surroundings and appears darker, producing a "blemish" on the surface. Although some spots appear very dark, this is a false impression, because even the darkest sunspot, if seen alone in the sky, would still be about as bright as the full Moon (Figure 1.5).

Directly above the photosphere is the chromosphere, and beyond that the corona. These two features are easily visible during a total solar eclipse, the chromosphere as a reddish-pink ring hugging the limb of the Sun at totality and the corona as the ghostly white solar atmosphere extending several radii beyond the eclipsed Sun. Nearly all the activity on the photosphere, in the chromosphere, and in the corona is related to the Sun's magnetic field.

Chromosphere and Corona

Beyond the photosphere is a layer called the chromosphere (*chromos* means "color"). With a thickness of about 2000 km, the temperature in the chromosphere is typically 10,000 K. The gas of the chromosphere is rarefied, or less dense than below, making it difficult to see because of the overwhelming intensity of the photosphere. The chromosphere is reddish-pink because of its emission strength at 656.3 μm, which is the wavelength of the Hydrogen-alpha line in the solar spectrum. By utilizing special instruments that pass only certain wavelengths of light, astronomers are able to study features within the chromosphere.

Activity in the chromosphere takes a variety of forms. One large-scale feature is the chromospheric network. This is a net-like pattern that overlays the super-granulation pattern visible in the photosphere. The network is made of small patch-like areas only a few arc seconds across. The patchwork appears bright when viewed in the light of Calcium. If the network is viewed in Hydrogen-alpha light, dark protrusions called mottles are revealed. When the mottles are longer than a couple seconds of arc, they are known as fibrils.

Spicules are easily seen at the limb as an emission (bright) feature appearing like tiny jets of gas flaming out of the Sun. On the disc they appear dark, taking several forms, such as "brushes" or "chains." Spicules rise to an average height of 7500 km and a width of about 800 km.

Since hot gas is ionized, it clings to the strong magnetic fields of the Sun, tracing out lines and loops where they exist. This allows us to see the shape of a magnetic field, especially when it is silhouetted against the darker background sky at the solar limb.

Prominences are clouds or stream-like projections of gas visible above the solar limb. Seen against the disc of the Sun, prominences appear dark and are called filaments. Although prominences come in a variety of shapes and sizes, they tend to fall within two basic classifications, active or quiescent. Active prominences have shorter lives and sub-class names such as surge, spray, jet, or loop. These are energetic events that sometimes end with the prominence being ejected into the corona or beyond. Quiescent prominences last longer, appearing at times on the limb of the Sun as a mound or hill. They are static and slow to change appearance. Prominences are suspended above the photosphere by magnetic fields. Physically, they can be a few thousand to several hundred thousand kilometers in length, ten thousand or more kilometers in width and height, and around 10,000 K in temperature.

Another chromospheric curiosity is the solar flare. It is believed that flares result from released stress within the magnetic field of a sunspot group. Flares sometimes appear as a sudden brightening of an existing plage (bright patchy region within the chromosphere). The initial phase of brightening can be rather quick, from a few minutes to an hour. A gradual decline in intensity of the flare is experienced following peak brightness. The energy output of a flare can be truly astronomical. It would not be unusual for a large flare to produce the equivalent of several seconds' worth of the Sun's total output. Pack all this in an area less than one hundredth of a percent of the surface area of the Sun, and the outcome is spectacular.

Solar flares are known to eject particles of matter from the Sun in addition to deadly radiation. Within a matter of hours to days these particles can reach Earth, disrupting communications, power grids, or damaging spacecraft. This is why it is important for scientists to track solar flare activity around the clock.

Occasionally, an event called a coronal mass ejection, or CME, takes place. The coronal mass ejection is an expulsion of a part of the corona and particles into interplanetary space. CMEs can represent the loss of several billion tons of matter from the Sun. These particles can move at velocities near 400 km/s. Solar flares appear to trigger some CMEs; other CMEs occur without an accompanying flare. The ejected particles are carried by the solar wind to our vicinity in space, causing havoc and initiating beautiful auroras in the polar regions of our planet.

Extending further out from the chromosphere, the gas becomes particularly tenuous. There is a thin region between the chromosphere and corona called the transition zone. From this point outward the temperature begins to increase markedly. Within the corona, to a distance of several million kilometers from the photosphere, temperatures exceed 500,000 K and at times may be more than 2,000,000 K. The heating mechanism of the corona is unknown and remains one of the big questions for solar astronomers to answer.

The light of the corona (meaning "crown") compared to the photosphere is extremely weak and can only be seen during a total eclipse of the Sun or by using a special instrument called a coronagraph, which creates an artificial eclipse. Spacecraft that get above Earth's atmosphere have a particular advantage in studying the corona. Scattered light from dust, refraction, and water vapor in the atmosphere hamper the Earth-bound observer, but in space these conditions do not exist, favoring observation of the Sun's corona.

The shape of the corona varies with the strength of the sunspot cycle. During sunspot minimum the corona is seen more fully and is extended around the solar disc. At sunspot maximum the corona is restricted to the equatorial or sunspot zones. This restriction is attributed to an increase in magnetic activity during the solar cycle.

As the corona reaches further from the Sun, it eventually becomes one with the streams of charged particles escaping the Sun, called the solar wind. Comets are excellent proof of the solar wind's existence. As a comet approaches the inner Solar System material in it is stirred up and gassed out by the heat experienced as it nears the Sun; the solar wind then pushes this material away, forming the comet's tail.

Energy begins as hydrogen in the Sun's core, and through the proton-proton cycle this energy becomes photons and particles slowly making their way through the solar interior to the photosphere, where they are released, giving us warmth, light, and life. In a way, we are residents of the Sun, living in its stream of constant emissions. Having such a close relationship with a natural occurring power plant, we are much the wiser to strive for a full understanding of how the Sun affects Earth, and ourselves.

The Earth-Sun Relationship

Within the last decade or two a new term, space weather, has come into vogue to describe the environment of space near Earth, as it has been affected by the release of energy and particles from the Sun. The study of space weather is critical in understanding Earth's environment.

Earth, with its iron core, behaves as though it were a huge magnet, having north and south poles due to the dynamo effect. Lines of magnetism emerge from these poles and arch out into space for tens of thousands of kilometers and return to the opposite pole. Although we usually think of magnets as possessing attracting power, repulsion is also a characteristic of magnetism. Objects that have a charge and are moving will be repelled, or pushed away, by a magnetic field. The large magnetic field surrounding and protecting Earth is called the magnetosphere.

As discussed earlier, the Sun has a steady output of charged particles and bits of matter that is collectively termed the solar wind. Streaming throughout the Solar System at about 400 km/s the solar wind flattens Earth's magnetosphere on the side facing the Sun. It is as if you were walking against a strong breeze, with an

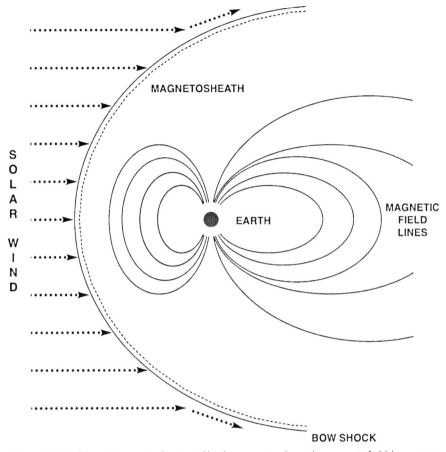

Figure 1.6. Earth's environment is dominated by the magnetosphere, the magnetic field that protects us from the ill effects of the solar wind.

umbrella pointing out in front of you. The umbrella repels the wind around you, while the force of the wind is felt pushing on the umbrella's fabric. The magneto-sphere can be pictured similarly repelling the charged particles from the solar wind around Earth and providing protection from it (Figures 1.6 and 1.7).

Weather patterns on Earth give rise to hurricanes, tornadoes, and thunder-storms. The Sun likewise has its "weather pattern" that powers massive solar flares and CMEs. During a solar flare or CME, the rapid ejection into space of subatomic particles and a large amount of X-ray, UV, and gamma ray radiation increases the volume of particles reaching Earth. As these events happen, particles often get caught up in our protective shield, disrupting the magnetosphere and affecting adversely the local space weather.

Sometimes damage can result to national power grid structures. The delicate electronics housed in satellites above the atmosphere are especially vulnerable to electrically charged particles. Some satellites have been put in jeopardy because a

Figure 1.7. The aurora during active solar periods can be visible from middle latitudes. Illustrating the effects of charged solar particles on upper atmospheric gases, this auroral display near 40° north latitude was observed November 7, 2004. Lex Lane.

solar storm placed additional drag on the satellite and reconfigured its orbit. Radio, TV, and telephone transmissions may be interrupted; other aspects of Earth's atmosphere can also be altered, creating holes in the ozone layer. These and other realities have been known for many years.

Auroras (borealis in the north and australis in the south) are sometimes visible even in the lower latitudes following a particularly energetic solar event. Gusts in the solar wind resulting from flare or CME activity excite the magnetosphere of Earth, producing electrical currents that travel through the field lines leading to the magnetic poles. Gases in the upper atmosphere will then have their atoms strengthened to higher energies. When these atoms release their energies as photons, the gases glow, creating the beautiful aurora.

Just as a meteorologist studies and forecasts weather on Earth, astronomers and physicists work to keep watch on the near-Earth space environment. Making space weather forecasts, and providing useful solar-terrestrial information, is the domain of the National Oceanographic and Atmospheric Administration (NOAA). The NOAA and the U. S. Air Force cooperate in a joint effort overseeing the Space Environment Center (SEC), which monitors and distributes space environment information in a timely manner.

Much of what we have learned has been realized in only the last several decades. Solar space missions such as NASA's Transition Region and Explorer Spacecraft (TRACE) and the Solar and Heliospheric Observatory (SOHO) have provided answers to many questions. Of course, for each answered question more are begging to be asked. Some structures of sunspots are yet to be understood, and there are river-like flows of plasma in the polar regions that require comprehending. And what causes the irregularity of the 11-year solar cycle?

We struggle to see inside distant stars, which are far beyond our reach. The processes of those stars we partly understand because the Sun has provided a laboratory right in our backyard for the study of its multifaceted behaviors. For the suitably equipped amateur astronomer, the Sun provides a lifetime of observing enjoyment. Explore the options presented in the following pages and discover that the choices of study are as varied as the spotted face of the Sun.

Chapter 2

Some Basics
of Solar Observing

Why Observe the Sun?

Charles A. Young (1834–1908), a professor of astronomy at Princeton University and a well known solar observer, in 1901 wrote that "the Sun is a star, the nearest of them; a hot, self-luminous globe, enormous as compared with the Earth and Moon, though probably only of medium size as a star; but to the Earth and the other planets which circle around it, it is the grandest and most important of all the heavenly bodies." Professor Young was a pioneer in solar physics. Instrumental in several new fields of study, he was the first to photograph a prominence, he discovered the chromosphere's reversing layer, and he was a popular public speaker on scientific matters. To Young, the Sun *was* the most important of all the heavenly bodies, his above statement concluding with the line, ". . . and its rays supply the energy which maintains every form of activity upon their surfaces." He was speaking of the surfaces of the planets, particularly Earth, with its flourishing plant and animal life, weather systems, oceans, and rich atmosphere. Without the Sun, none of this would be possible.

Because of the Sun's nearness (150 million kilometers away) and size (1.4 million kilometers in diameter), it provides us with a unique opportunity for detailed studies of a star. Nearly all other stars are seen as points of light, in even the world's largest telescopes. Indirectly, spectroscopy and astrometry provide rich data about stars, but still these techniques furnish only a glimpse of what could be learned from direct observation.

Amazingly, with only the unaided eye and a total eclipse, we are close enough to learn that the Sun has a shape-changing ghostly white atmosphere and beautiful pink prominences. Early Chinese astronomers viewing it when close to the horizon or through thin clouds were able to discover that the Sun develops dark spots. That was not the correct way to view the Sun, but with proper eye protection we, too, can see naked-eye sunspots that occasionally appear on the solar disc. An enduring large sunspot would demonstrate that the Sun rotates, and additional observations would allow for the determination of an approximate rotation rate. Therefore, our nearness to the Sun has been key in learning how an average star behaves.

When a telescope with special viewing apparatus is used to examine the Sun, the grandeur spoken of by Young can be breathtaking. Activity on the Sun is the most energetic in our Solar System. The amateur observer can expect to follow the daily growth of individual sunspots and sunspot groups. Prominences erupt, and then

J.L. Jenkins, *The Sun and How to Observe It*, DOI 10.1007/978-0-387-09498-4_2,
© Springer Science+Business Media, LLC 2009

Figure 2.1. Solar observers must always strive to view the Sun in a safe manner.

rain or fall back onto the surface below. Solar flares, usually only seen with special filters, may brighten in a few minutes to such intensity that they become visible in white light. It is truly awe-inspiring to witness in a single event the release of energy equivalent to the total output of humankind's throughout all of our history. In what other astronomical pursuit can such an abundance of spectacular activity occur in so short a time span?

A number of solar observers were asked why they found the Sun an appealing target for their telescopes. Many, of course, responded by making reference to the solar dynamics just mentioned, but several spoke of Sun observing as being a humbling and relaxing experience. One observer quipped humorously that the Sun was easy to find in the daylight, and another said that it was simply fun to share with the public. Fun and the fascination of examining a star close up and in detail seem to be a common theme among die-hard solar observers. It is for these reasons so many amateur astronomers study the Sun.

Although solar observing is attractive to the amateur astronomer it does present a few stumbling blocks that a night sky observer doesn't face. These obstacles include the three "S's": **S**afety, **S**eeing, and **S**uitable instrumentation. Barriers, indeed, but easily overcome when know-how, determination, and proper equipment are all applied to the challenge (Figure 2.1).

Safety and the Sun

The first lesson that must be learned by all astronomers wishing to observe the Sun is of the inherent danger involved in this activity. **NEVER attempt to view the Sun through any optical instrument that has not been properly fitted with SAFE solar observing appliances. NEVER stare at the Sun with your unaided eyes, unless looking through a known and tested solar filter intended for such use.** It only takes the briefest amount of exposure to unfiltered sunlight to permanently damage or destroy your eyesight. In this activity you're given only *one* chance, so be cautious and *never* take the Sun for granted.

To demonstrate the potential risks involved, perform this simple experiment. Point a telescope toward the Sun with no solar filters in place; a refractor is best for this illustration. Remove the eyepiece and star diagonal. Then, while standing to the side of the telescope, away from the beam of light, hold a sheet of paper at the focal plane of the objective. What is the result? Of course, after several moments the paper will be ignited (be careful here)! This should be sufficient in demonstrating the amount of heat present at the focus of a telescope NOT properly fitted for safe solar observing.

Although the threat from heat is obvious, safe harmless solar observing is done everyday, by observers using equipment available to even the novice. The key to risk-free solar observing is to become educated on what is safe and what is not. In that way, potential problems are avoided.

The least expensive and safest method of viewing the Sun's photosphere is the projection method. A refracting telescope or Newtonian reflector is suitable for solar projection. Either of these instruments can easily support the mounting of a projection screen, while their simplicity offers little opportunity for damage to the telescope from heat. Do not choose compound telescopes, such as the popular Schmidt-Cassegrain, for solar projection. Heat damage to internal components, like a plastic baffle tube or the secondary mirror, is a possibility with a Schmidt or similarly designed telescope.

To project an image of the Sun onto a white screen, a Huygens or Ramsden eyepiece is used, with a maximum telescopic aperture of about 100 mm (4 in). Larger aperture telescopes risk focusing too much concentrated heat at the focal plane, where the eyepiece is located. Some observers prefer using a smaller (80 mm or less) aperture telescope with a higher quality eyepiece that gives a crisper appearing projection. An Orthoscopic or Plossl does well with a small telescope, but take care; the outside chance of melting the optical cements used between eye lenses does exist. The white viewing screen is positioned some distance behind the eyepiece of the telescope pointed at the Sun. This arrangement enables features of the white light Sun to be easily seen. The greater the distance separating the eyepiece and the screen, the larger but dimmer is the projected image. So, a compromise between size and brightness of the projected image is a serious consideration when projecting the Sun through a telescope.

For white light observations, an alternative to solar projection and the means most often utilized by amateur astronomers is the direct view. Objective filters fit over the entrance of a telescope, reducing the brightness of the Sun and screening out the harmful light rays that otherwise would reach the observer's eyes through the telescope. These types of filters are available in several forms, from the specially coated glass and metal models to those made of optical grade mylar. Most filter manufacturers sell pre-mounted solar filters that fit commercial telescopes, or in a cell to fit your specific telescope tube diameter. The filter product itself is often available separately if you wish to mount it in a homemade cell yourself. Next to solar projection, an objective filter is the safest method of viewing the Sun for today's amateur astronomer. More will be said regarding solar projection and direct viewing in the chapter dealing with instrumentation for the white light observer, later in the book.

Specifically, what are the dangers to avoid when observing the Sun? One thing is overexposure to light, primarily in the blue-green regions of the spectrum. This causes damage to the light sensitive cells of the eye called rods and cones. Chemical changes occur when we experience overexposure to bright sunlight, changes that leave us with blindness in our eyes either for a short period of time or permanently. This chemical retinal damage can result from either a single exposure event or the cumulative effects from a number of unprotected "short looks" at the Sun.

Although overexposure to the bright visible light of the Sun is a factor in blinding the observer, there are also other damaging rays. The Sun emits invisible radiation in the form of ultraviolet (UV) and infrared (IR) light. Ultraviolet light (280–380 nm), although not reaching the retina due to absorption by the eye, does contribute to the development of cataracts, as well as speeding up the aging process of the outer layers of our eyes. When infrared light (780–1400 nm) is admitted to the eyes, heat burns the exposed internal tissues, resulting in destroyed rods and cones. Much of the thermal energy from the Sun is in the form of IR light. A permanently blind region on the retina will be the end effect of this exposure. Unfortunately, the observer may not immediately be aware of the injury, as there is no means of sensing heat in the eyes. The effects are sometimes not *visible* until after the damage has been done, usually within a few hours (Figure 2.2).

Items that have sadly been used independently in the past and are NOT safe for solar observing include deposits of candle soot, commonly called smoked glass, polarizing and neutral density filters, exposed and developed color film, compact discs (CDs), silver-less film, and aluminized food wrappers. Many of these may appear to help dim the visible sunlight, but these items transmit high levels of IR light. For your own safety when observing the Sun, only use products intended for safe solar viewing.

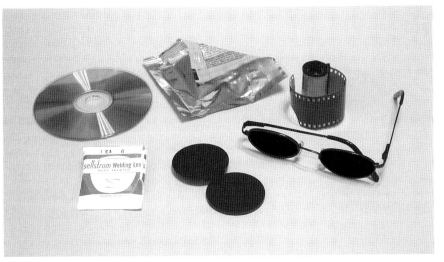

Figure 2.2. Items NOT safe for viewing the Sun through include CD's or DVD's, mylar food packaging, photographic color film, sunglasses, crossed polarization or neutral density filters, and welder's glass filter with a density less than shade #14.

A dangerous product that shows up even today, particularly with low-end telescopes in the used marketplace, is the eyepiece solar filter. These are filters intended to screw into the barrel of an eyepiece to facilitate direct viewing of the Sun. The filter is no more than a very dense-colored glass. When placed at the focal plane of a telescope pointed toward the Sun, the filter will eventually shatter from the concentrated heat, exposing the eye to the blinding light and heat of the Sun.

I will never forget the awakening experience I had of pointing my first telescope fitted with one of these filters toward the Sun. After stepping away for a few moments to retrieve my notebook, then hearing the ominous snap of cracking glass, I returned to find a beam of white sunlight streaming out of the eyepiece. A few moments earlier or later with the accident, and I could have been blind in one eye! For the safety of yourself and others discard these filters immediately.

So, what constitutes a safe white light solar filter? To be a visually safe device, the appliance must reduce the intensity of the light from the Sun entering the telescope (280–1400 nm) to a level of 0.003% (density of about 4.5).[1] This is the *lower limit* of safety; many individuals prefer a filter transmitting a bit less, typically a filter with a density of about 5.0 is found to be comfortable by most observers. At this level, the Sun appears about as bright as the full Moon.

The manufacturer of your filtering appliance must always have the final word in how it is to be used in solar observing. For example, some manufacturers of white light filters supply what is termed "photographic density" versions of their solar filters. Warnings from the manufacturers are often given concerning the visual use of these filters; these are warnings provided with safety in mind. The purpose of photo-density filters is to shorten exposure times when photographs are made of the Sun. Because photographic density filters transmit more sunlight (including UV and IR) than regular visual filters, they are *not* intended for anything, other than photographic use.

Occasionally, a large sunspot will appear on the solar disc, and the novelty of discovering whether you can spot it with your unaided eyes becomes a challenge. A piece of shade #14 welder's glass of a size suitable to cover both eyes is a superb filter for naked-eye observation of the Sun. Available at welding supply stores in rectangular- and disc-shaped pieces of several sizes, these filters sufficiently diminish the intensity of the sunlight in the visible, UV, and IR wavelengths. Due to their poor optical quality, however, they are not recommended for use over the telescope objective, and never at the eyepiece as the sole means of light reduction (Figure 2.3).

So far we've only been discussing white light filters, but monochromatic observers must also be cautious with safety issues. The H-alpha and Calcium-K line filters widely available to today's amateur astronomer are totally safe, if used according to the manufacturer's instructions. A specially made pre-filter called an energy rejection filter (ERF) can be used over the telescope's objective with some systems to absorb or reject UV and IR light. The narrow band filtering unit itself contains additional trimming and blocking filters that remove unwanted off-band wavelengths, with the end result being a safe, monochromatic view.

Always check and double check the filtering systems on a telescope, making certain that the components are properly and securely mounted. Eye-end filters can be knocked off a telescope and objective filters can become dislodged by the wind. Avoiding a perilous accident in those circumstances is just a matter of luck, and you should never leave your safety to luck.

Another issue an observer of the Sun should consider is how to find the target in the field of the telescope. An excellent habit to develop is the removal or capping of the main instrument's finder telescope before solar observing. An uncapped finder becomes a small projection telescope just waiting to burn a shirtsleeve or an arm

Figure 2.3. The #14 shade of welder's glass makes a wonderful filter for safe unaided viewing of sunspots or the partial phases of a solar eclipse.

Figure 2.4. Use a pinhole device, such as the commercial unit shown, or watch the shadow of the telescope on the ground to center the Sun in the telescope's field of view.

that is accidentally placed in its light path. Also, never sight along the edge of a telescope tube for the purpose of locating the Sun in the telescope. That is the same as staring at the Sun with your unaided eyes!

You can, and most solar observers do, watch the shadow formed on the ground by the telescope, as it's directed toward the Sun. The telescope's shadow will shorten and form a round circle when the Sun is within, or very near, the field of view. There are several manufacturers of simple "pinhole solar finders" that attach to a telescope tube, allowing accurate placement of the Sun in the field of view. One made by a popular manufacturer is mounted next to the regular telescope finder. Many craft-minded observers have constructed their own solar locating devices based on these same principles, and so could you (Figure 2.4).

The Sun is tamable. The keys to safety are staying informed and always following proper procedures. Make a conscious effort to not become a statistic, other than as a safe solar observer. Remember, have respect for the Sun, and you will always be able to observe safely its unique, ever-changing features.

Seeing Conditions

At a local astronomical event several years ago, I was part of an informal discussion in which the topic turned to the amount of detail one observer could expect to see on the Sun with his telescope. Of course, we talked about the aperture of the telescope, filters that could improve the appearance of some features, and the advantages of solar observing during some parts of the day rather than others. One long-time solar observer was present and made a statement which rang true. In response to what the amateur astronomer could expect to view through his telescope he said, "All things considered, seeing is everything!" This is the Holy Grail of solar observing.

Astronomical seeing is defined as the quality of the atmosphere between the observer and what he or she is viewing. It is the state of the medium through which waves of light must transverse before they enter your eyes. "Daytime seeing," as it is commonly called, creates circumstances where the Sun may appear perfectly still and richly detailed or other times when all detail is awash, smeared to a point where only large sunspot umbra are discernible.

When the seeing is excellent, and using a 125 mm or larger telescope, photospheric granulation will be clearly visible, penumbral filaments (dark thread-like structures radiating from a sunspot umbra) are well defined, and pores (tiny black spots) appear steady, not popping in and out of view. Unfortunately, these moments are rare. More times than not, daytime seeing conditions will be in the range of somewhat less than perfect to totally unstable. Being a newbie to astronomy, I was hardly conscious of astronomical seeing conditions. My first telescope was of small aperture and rarely used in excess of 50× magnification. But as I became more accustomed to observing and my "astronomical eye" developed, I also perfected an appreciation of observations made through a steady atmosphere. Whether a beginner or advanced amateur astronomer, it pays to make the effort to understand how the atmosphere influences our observing experiences. Why? Because seeing really is everything (Figure 2.5).

The nighttime observer is aware that stars rapidly twinkle and fluctuate with brightness. This is called scintillation and occurs because of turbulent conditions in Earth's upper atmosphere at heights of 2 or more kilometers affecting point sources of light, such as stars. A swelling, blurring, or general degrading of the telescopic view marks yet another kind of seeing, called image motion. Degradation of a

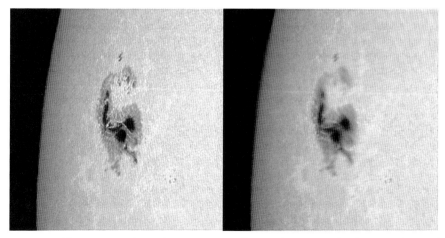

Figure 2.5. These images obtained moments apart illustrate the range in local seeing. The good conditions seen on the *left* allow fine detail to be visible. Fine detail is lost with poor seeing on the *right.* AR0808 from September 9, 2005, at 1419 Universal Time. Jamey Jenkins.

celestial object due to motion can be anything from a slow, lazy, rolling of the view to a rapid boiling. Image motions seem to originate in the lowest levels of the atmosphere, including the immediate vicinity of the observer or inside the telescope itself. The layer of air closest to Earth, the troposphere, is responsible for our daily weather patterns as well as the temperature variants that are at the root of most poor seeing conditions. During the day there are fluctuations in the temperature of the air caused partly by the heating of ground objects and the re-radiating of those thermals into the local surrounding air, that is, the lowest 100 m. It is these low-level convection thermals we try to understand and control when we are trying to reduce the poor seeing conditions associated with solar observing.

So, what does a solar observer do to combat poor daytime seeing conditions? First, evaluate your observing site. Do buildings sit beneath or near the path that light from the Sun must take in reaching your telescope? If so, relocate and avoid that situation. If possible, locate the observing site in a place where the prevailing winds are not blocked. Trees, buildings, tall fences, hills, and other obstructions can cause disturbances in local airflow patterns, resulting in poor seeing.

Amateurs with an observatory find that a building designed with a roll-off or split roof is preferable to a dome. Domes tend to create and preserve poor seeing conditions because the area around the entrance slit re-radiates thermals. Also, air currents originating within the dome and exiting the slit contribute to poor seeing. Building materials for the observatory are selected to be as lightweight as practical, so as not to store up large amounts of heat. For the individual without a permanent observatory, an open-air site is superior to observing through a window or garage

Figure 2.6. A suitable alternative to domed observatories, which tend to offer poor daytime seeing, is the roll-off roof design.

door opening. Again, avoid large heat sinks such as asphalt or concrete parking lots; a large grassy area where airflow is not restricted would be ideal (Figure 2.6).

Except for the interior of the telescope, avoid flat black surfaces. White paint reflects heat, while black absorbs it, re-radiating the heat back into the light path of your telescope. This could be extended to include the surrounding observatory structure and auxiliary equipment. If constructing your own solar telescope, use a tube size a bit larger in diameter than necessary, so as to keep sunlight from striking the sidewall of the telescope tube, creating harmful internal air currents.

Over time, study the local seeing conditions and weather patterns, and note when prime solar observing occurs for you. Is the air steadier after a cold or warm front has passed? Is there a part of the day when the Sun appears less turbulent to you than other times? Many experienced observers find early morning to be their prime solar observing time. This is the time of day before the Sun has had a chance to heat local rooftops, driveways, and structures. Since my eastern skyline is restricted, preventing early morning viewing, my personal preference has been nearer to mid-day, with the Sun high in the sky and a slight east-west breeze in the air. This has proven to be a successful combination for me. Why? Because with the Sun near the meridian, the light path is shorter through the most turbulent layers of atmosphere, and the gentle breeze reduces any rising thermal currents near the telescope. There is no rule of thumb for the time of day that is best for solar observing. What is true is that daytime seeing varies with the conditions present at any given site. Study those conditions during superior seeing, and when they repeat, take advantage of the opportunity.

Heat is not the only deterrent to good seeing conditions. Atmospheric refraction gives rise to a deterioration of the telescopic view because of a prismatic effect resulting from the curvature of Earth's atmosphere. Have you watched the planet Venus rise or set in the morning or evening sky through a telescope? Probably as the planet draws near the horizon you've taken notice of the accompanying tiny rainbow. The further away from the horizon, the less prominent is the prismatic effect, but when an object is only 25° from the zenith, the atmospheric refraction is still greater than 1 arc second.

To avoid atmospheric refraction, solar observers filter the telescopic view, removing all the rainbow's colors but one. This can be accomplished by using standard colored eyepiece filters in addition to your normal white-light filtering appliance. Monochromatic observers have essentially removed the effect, because their narrow band equipment passes only a thin slice of visible light from the solar spectrum. Colored glass filters, although lessening the effects of atmospheric refraction, are also used to enhance selected white-light features on the Sun. For example, green filtration increases the contrast of photospheric granulation and faculae. The chapter in this book on white light observing techniques will explore in greater detail filter use; for now realize that the narrower the bandwidth of a filter, the lower in the sky you can expect to remove atmospheric refraction.

For record-keeping purposes, it is important to devise a uniform system that describes the seeing conditions a solar observer experiences. The difference between excellent and poor seeing is obvious. Excellent seeing conditions will permit observing the finest details for extended time periods. Poor conditions prohibit observing fine detail at all, or allow only a few details to be seen for short amounts of time. One method of judging the quality of seeing incorporates the visibility of a specific feature, granulation, as a measuring stick. By assigning a numerical value to the appearance of granules visible in the telescope, a scale

Table 2.1. A descriptive scale for defining seeing conditions. Many observers use either arc seconds or a "common" terminology to describe the seeing they experience during an observing session. Since common terms tend to be subjective, the arc second method is preferred among experienced observers. Solar granulation and pores are the defining features of this scale

Description	Scale in arc seconds	Common term
Granulation resolved	<1 arc second	Excellent
Granulation appears blotchy	1–2 arc seconds	Good
Granulation/pores sometimes visible	3–5 arc seconds	Fair
Granulation/pores not visible, ill-defined sunspots	>5 arc seconds	Poor

representing the seeing conditions can be obtained. For instance, seeing that is better than 1 arc second means that granulation is seen clearly. A 1 or 2 arc second seeing is interpreted as the granulation appears mottled. Seeing in the 3–5 arc second range would occasionally allow a mottled view of granulation, and the tiniest sunspots or pores would appear to jump in and out of view. The arc second description is the preferred choice of many serious observers, providing a clear understanding of daytime seeing conditions.

There is another characteristic of atmospheric quality to be considered by the solar observer, termed transparency. Reduced by water vapor, dust, smoke, and other atmospheric particles, transparency is a description of the clearness of the sky. If transparency is sufficiently diminished, some features begin to lose clarity. At one extreme we find completely opaque clouds, through which nothing is visible. In the other direction, the sky will be a deep blue in the Sun's vicinity. Most days will be somewhere in between, and fairly subjective terms such as excellent, good, fair, and poor can be used to characterize transparency. Excellent describes the clearest conditions, good is clear but not as perfect as excellent, fair represents a hazy sky, and poor is thin clouds between you and the Sun (Table 2.1).

Reference

1. *Solar Observing Techniques*, C. Kitchin, Springer-Verlag, 2002

Chapter 3

Observing the White Light Sun

A dedicated solar telescope is a unique tool intended for viewing a single object in the sky, the Sun. Unlike a nighttime scope, an instrument for Sun viewing isn't expected to gather a lot of light. In fact, when observing the Sun most of the effort is spent in reducing the amount of light received. While a big "light bucket" can show you galaxies many millions of light years away, chances are on many days a small refractor will outperform it on our nearest star. Because telescopes for solar observing are usually 150 mm or less in aperture, the emphasis for the stargazer now switches from quantity to quality of light. Serious solar observers make use of instrumentation designed to permit optimal viewing of their subject, the Sun.

Telescopes for White Light Solar Observing

The Sun as it appears to the unaided eye is said to give a "white light" view. White light is the consequence of integrating all the colors of the visible spectrum, from violet to red. When we observe the Sun in white light, the layer we see is the photosphere. Try as we might, seeing below the photosphere in visible light is impossible – the gas is too dense and opaque. Above the photosphere the chromosphere becomes thin and tenuous; shining in the light of hydrogen it is far too dim to be glimpsed within the overpowering white light from below.

In theory, any telescope can be adapted for white light solar observing. The thing to remember is this: with the Sun, some telescopes perform better than others. Many solar observers come from a night sky background; consequently, their equipment is designed to function best in that environment. Adapting a nighttime telescope to useful daytime observing can be as simple as attaching a white light objective filter to the telescope. But if you want superior performance, construction of a dedicated telescope might appeal to you. My latest telescope, a 125 mm aperture f/18 refractor, was assembled with solar observing in mind. The focal length was selected to provide a specific size of solar image at the focus of the objective. The tube interior was carefully baffled to reduce scattered light, and the exterior was painted white to reflect heat. A few companies manufacture dedicated white light telescopes intended specifically for observing the Sun.

Many white light solar features are of low contrast. That is, there is only a slight difference in the brightness between the feature of interest and the nearby solar background. There are a few exceptions to this. Examples of contrasting features include a sunspot umbra seen against the surrounding photosphere or a brilliant

J.L. Jenkins, *The Sun and How to Observe It*, DOI 10.1007/978-0-387-09498-4_3, light bridge that crosses a dark umbra. To see detail within the penumbra of a sunspot or to study solar granulation, both features of which have low contrast, requires an instrument capable of producing a sharp, contrasting image with ample resolution. Therein are the keys to what a telescope intended for solar observation should be expected to yield: sharpness, contrast, and the resolution capable of defining the selected feature under observation. Keep these three criteria in mind whenever selecting a telescope intended for serious solar investigations.

Many amateur studies use a resolution on the order of 1 arc second, thereby permitting the observation of fine penumbral detail and granules. A 125 mm aperture telescope exhibits a theoretical resolution (Dawes' limit) of 1 arc second, and with fine seeing conditions, this would represent the minimum desired aperture for a serious solar observer. Greater aperture telescopes up to about 300 mm can be used successfully, though resolution for the most part becomes limited by the seeing conditions. Smaller aperture telescopes (less than 125 mm) take in reduced amounts of the disturbing atmospheric cells, making for steadier views, but a smaller aperture always results in a compromising of theoretical resolution. These smaller telescopes (50–100 mm aperture) are suitable for casual observing or making sunspot counts. Larger aperture telescopes will always show finer detail than smaller aperture instruments; what will be required is patience on the part of the observer for choice moments of steady seeing conditions.

Reflecting Telescopes

The simplest and the most often employed of the reflecting telescopes is the Newtonian, an invention of Sir Isaac Newton in the seventeenth century. A home-made or commercially manufactured Newtonian is useful for direct observation of the Sun in white light with the addition of a full aperture solar filter. Typical in the amateur ranks are the 150 mm (6-inch) Newtonians. These telescopes are reasonably priced, owing to the single optical surface of the primary mirror, a definite advantage for the cost-conscious hobbyist. Dollar for dollar the purchaser obtains more aperture with a Newtonian than with any other telescope design. Compound telescopes, those possessing mirror and lens primary elements and high-end refracting telescopes, are expensive partially because of the greater number of optical surfaces to be worked.

Newton's first reflector was less than perfect. It was made with a primary mirror suffering from a defect known as spherical aberration. A mirror or lens suffering from the aberration is unable to focus all the gathered light rays in a single plane. Said another way, rays that reflect from the objective farther off axis than others reach focus at a different point along that axis. To correct spherical aberration in a Newtonian telescope, the curve of the primary mirror is deepened from a sphere to the shape of a parabola. A parabolic mirror brings the light rays to a single point of focus.

Another common defect of the Newtonian is coma. A mirror with a short focal ratio creates an off-axis image of a point source that resembles a tiny wing or a V-shaped "smudge." The farther off-axis the light, the more pronounced is the aberration. One way that coma can be eliminated, or at least reduced, is by the use of special correcting lenses inserted within the light path. But the simplest solution is to use a moderate to large focal ratio (f/8-12) mirror in the telescope, creating a long focal length that is well suited for lunar, planetary, or solar observing.

The Newtonian telescope has a distinct advantage over a lens-based telescope in that it is perfectly achromatic – that is, without color defects. Color or chromatic aberrations result from the prismatic effect of refraction. Often a violet ring will be seen surrounding a bright star viewed through a refracting telescope. Mirrors, since they operate via reflection rather than refraction, are free of color aberrations.

An amateur astronomer armed with a well-made 150 mm Newtonian and full aperture solar filter can expect to observe all the interesting white light features of the Sun. These include limb darkening, the umbral and penumbral details of sunspots, pores, light bridges, faculae, and granulation. Somewhat larger aperture instruments, though less portable and susceptible to poor seeing, can also be used if the objective filter is placed off-axis (OA). For the Newtonian, an OA filter is circular and big enough to fit between the vanes of the diagonal mirror support while being contained within the limits of the primary mirror. Expect an OA filter to increase slightly the contrast lost from a secondary mirror and support system being located within the telescope's light path. However, with a 150 mm or smaller reflector, the off-axis arrangement won't work well. Resolution is too seriously compromised if a smaller (50–70 mm) off-axis aperture is used with that size telescope. Therefore, if you have a 150 mm or smaller telescope, use a full aperture solar filter; if larger than 150 mm use the full aperture or consider an OA filter as large as possible (Figure 3.1).

Depending on the telescope, it may be necessary to mask or close the opening surrounding the primary mirror at the end of the Newtonian tube. Daylight leaking from the bottom of the tube upward onto the diagonal mirror can wash out the view seen in the eyepiece. A telescope tube having end rings provides an easy fix if you first cut a thin black cardboard disc having the same outside diameter as the tube. Next, remove the ring at the mirror end of the telescope, slip the cardboard disc into the ring, and snug-fit the ring back onto the telescope tube. Presto! The mirror end of the tube is masked off. Alternately, some sort of light blocking cap can be made or purchased to slip over the end of the tubing, doing the job admirably.

Figure 3.1. An off-axis objective filter is suitable for a Newtonian or catadioptric telescope. The full aperture objective filter on the right takes advantage of a telescope's resolution by utilizing the entire aperture of the telescope.

Other than direct observation of the Sun through a Newtonian equipped with an objective filter, white light observing can be accomplished via solar projection. This technique involves having the telescope serve as a "projector," forming an enlarged image of the Sun on a screen some distance from the eyepiece. The mounting of the shaded viewing screen to the side of a Newtonian's tube although possible can be frustrating, owing to the likely instability of such an arrangement. The open-ended tube design of the Newtonian also promotes internal air currents that are then magnified by heat from the Sun. These tube currents produce destructive seeing conditions, the nemesis of every solar observer. Heat buildup from solar projection that occurs at the secondary mirror of a large aperture telescope can damage or distort that optic, or at least create poor seeing. A few amateurs have successfully used a Newtonian for projection; however, it's the refracting telescope that provides the better choice when it comes to that technique.

In summary, the white light solar observer will find that a Newtonian telescope is the most cost effective. In the 150 mm or greater aperture using a full aperture solar filter, resolution is sufficient for most solar studies. Color aberrations are nonexistent in the Newtonian, and longer focus instruments are capable of providing superb images. For white light observing, the Newtonian can be an excellent solar telescope.

Catadioptric Telescopes

When a combination of mirrors and a corrector plate is used in a telescope, it is known as a catadioptric, or compound, telescope. There are two competing designs of compound telescopes commonly in use within the amateur ranks, the Schmidt and the Maksutov. Both are popular with many observers and have distinct advantages for the solar observer.

Master optician Bernard Schmidt invented the Schmidt telescope around 1930. Through the use of a spherical primary mirror and a uniquely shaped corrector lens at the front of the instrument, he succeeded in developing a coma free camera intended for photographing the night sky. There are a number of professional observatories stationed around the world that utilize Schmidt's highly successful creation in large survey-type instruments.

When modified to a Cassegrain configuration, the Schmidt became extremely popular with amateur astronomers during the 1970s and is still popular today. Portability, with a large aperture and long focal length in a compact package, is the greatest asset of this design. The amateur astronomer lacking space for a backyard observatory but desiring a sizable instrument finds the perfect solution in the Schmidt-Cassegrain telescope (SCT). For imaging purposes, the focal plane of a SCT is very accessible. Telescope manufacturers have also provided a wide array of accessories for these instruments. A downside to the SCT is the relatively large central obstruction (perhaps 30% of aperture), resulting in a slight loss of sharpness and contrast.

Dmitri Maksutov developed the Maksutov telescope in 1944. A Maksutov or "Mak" uses a spherical primary mirror and a meniscus lens at the telescope entrance to correct aberrations. A Mak with a Cassegrain configuration generally will have a longer focal length than a similar aperture SCT. An aluminized spot on the back side of the meniscus corrector often serves as the secondary mirror of the telescope. In most designs, the Mak secondary mirror is smaller than the secondary found on an identically sized SCT, resulting in views with slightly improved contrast. This feature makes the Mak a slightly better choice for solar observing. As with the SCT, portability is a distinct advantage. The relatively thick correcting lens (more than 12 mm) creates weight considerations plus the added expense of producing a large optic. These factors may limit the size of an instrument for the amateur.

As with the Newtonian, direct white light solar observing with a 150 mm or less aperture catadioptric telescope is best accomplished using a full aperture solar filter. Off-axis arrangements can be considered, if the telescope is large enough to permit a filter that doesn't limit the desired resolving power. Solar projection is *never* recommended with catadioptric telescopes. The potential for damaging or destroying the internal parts in a Mak or SCT is high. For instance, a plastic baffle tube can quickly melt and smolder due to excessive heat from the Sun or the secondary mirror can become overheated, distorted, cracked, or otherwise damaged. Even without these risks, the stubby fork mounts many of these telescopes use are not convenient for use with solar projection screens.

When it comes to general all-around astronomical observing, a catadioptric is likely the first choice of the amateur astronomer. Lunar, planetary, and solar observers appreciate the long focal length (f/10 or greater) and good image quality of the catadioptric. Deep-sky advocates recognize the large aperture of these telescopes, and the portability they offer is a definite plus. For direct viewing of the white light Sun, a catadioptric is a good choice of telescope.

Refracting Telescopes

A refracting telescope uses a lens as the primary optic to gather and focus light. Tradition gives spectacle maker Hans Lippershey of Holland credit for the invention of the refracting telescope, around the year 1608. Galileo Galilei turned his instrument skyward to make numerous discoveries that earlier non-telescopic astronomers were unable to see. By today's standards, his single-element objective telescope would be sub-par at best. Chromatic aberration, where colors fail to come together at a single focus in the lens was a weakness of the first telescopes. With time, the discovery was made that by combining glasses of different refractive properties a telescope maker could reduce or eliminate this problem.

Amateur astronomers today have two basic designs of refracting telescopes at their disposal, the achromat and the apochromat (APO). Achromat implies "colorless," and apochromat means "more colorless." A lens with two elements, usually of crown and flint glass, is referred to as an achromatic doublet. To perform well, an achromatic lens will have a long focal length, with a focal ratio of f/12-16 or even greater; shorter focal length doublet lenses suffer increasingly from chromatic and other aberrations. In the standard doublet refractor, long focal length minimizes color errors, although with the crown-flint glass combo, a trifle bit of color fringing may still be visible. One technique a number of astronomers use to remove this color is to observe through a yellow or yellow-green Wratten filter at the eyepiece. These filters effectively clean up the field of view by absorbing excess colors and transmitting only a single color, which comes to an accurate single focus. A few commercial sources provide interference type filters, sometimes called "minus-violet," that also remove color fringing in a refractor.

There are in the marketplace two element refractor lenses made with special extra-low dispersion optical glass, commonly called ED glass. A doublet objective constructed with ED glass is designed to precisely correct color across the visual spectrum. This improvement in color fringing over a conventional doublet lens is marked. The ED glass objective is considered apochromatic, since it is without color error. Other APO objective lenses may depend on three or four elements, one of which will certainly be made of calcium fluorite. APOs, whether of two, three, or four elements, are as free from color error as a lens system can be made. Because it is truly color-free, an APO telescope can be manufactured with a shorter focal length than the standard achromatic refractor, reducing the overall package length and making for a more portable telescope. However, due to the manufacturing costs of additional lens elements, an APO telescope is quite expensive.

Of the telescopes available, a refractor serves best for the projection method of observing. The refractor's straight through, closed tube design will minimize tube currents, and the instrument lends itself to the mounting of a shaded projection screen. Still, it is wise to consider the materials that have been used in the construction of your telescope. If a plastic drawtube or light baffles are used (as in some newer instruments), there exists a danger that heat from the Sun might damage those components. Older and high-end telescopes are usually of all metal construction and would be risk free. We recommend thoroughly inspecting any telescope before attempting solar projection, to determine if the components in the light path are combustible.

For the serious amateur astronomer observing the Sun, a refractor is the instrument of choice. Lacking the central obstruction of a mirror-type telescope, a similar aperture refractor will produce the sharper, more contrasting telescopic view. A full aperture solar filter easily mounts over the primary objective, creating a safe observing situation. And accessories for observing, such as a filar micrometer or camera, are readily attached to the refractor, supplying plenty of focusing room. Ideally, a well-made, medium focal length apochromat, with as large an aperture as is within your budget, would be your first choice. A mid-focal length telescope mounts sturdily, and the focal length may be bumped up as needed, with the addition of a precision Barlow or other amplifying lens. From the performance standpoint, few telescopes can compare to an apochromat for incredibly sharp, contrasting images. If cost is a factor, consider an ED glass doublet, or invest in the longer focal length traditional achromat.

Observing by Solar Projection

The oldest known successful technique for telescopic observation of the Sun is projection. An interesting engraving in the book, *Rosa Ursina sive Sol* from 1630, by Jesuit priest Christopher Scheiner, illustrates a refracting telescope being used for solar projection. The basic idea is to point the unfiltered telescope skyward, permitting it to create or project an image of the Sun onto a smooth white surface or screen some distance behind the eyepiece. Because the viewing of the Sun is accomplished indirectly with this technique, it is the safest method available. When used properly, there is no risk of eye damage from heat, brightness, or invisible forms of radiation. However, always be cautious of children or uninformed adults left alone with a projection telescope aimed at the Sun, particularly if there exists the possibility of someone inadvertently looking through the telescope. Never insert a hand or piece of clothing into the light beam emerging from near the projection eyepiece; if you do, you should expect to be burned!

A refracting telescope or Newtonian is particularly adaptable to solar projection. Between the two, the refractor is the more desirable for several reasons. Convenience of mounting the projection screen on a refractor is one; another is the improved quality of seeing experienced with the refracting telescope. Air currents found in a Newtonian exist within the refractor, but for the most part they settle down when the air temperature inside the telescope tube becomes stable. In an open-ended Newtonian, the air is constantly moving in and out of the tube, mixing with the outside cooler air. A catadioptric telescope should *never* be used for solar projection because of the risk of damaging the internal components of the telescope from the heat of the Sun (Figure 3.2).

Besides the safety factor provided by solar projection, a distinct advantage of this technique is that it is convenient for group observing. Usually the viewing screen will be positioned to show the whole disc of the Sun. Observers can be arranged around the screen, each with a clear view of the photosphere. For educational purposes, or casual observing of white light features, this technique works splendidly.

In order to see features clearly, projection is done onto a white Bristol board or card stock. The surface of the screen should be matte finished (to prevent glare), and slightly larger than the projected disc of the Sun. Pores and small features can become lost in the grain of paper; therefore, a very smooth surface is desired. The key to successful solar projection is providing a shaded environment for the projection screen. Indirect daylight falling on a projected image has the tendency to wash out all but the coarsest details. Contrast of a feature will be improved if the shade is used. Amateurs over the years have adapted various apparatus, from a wooden box to the interior of a building to create a shade. When solar observing consider using what is affectionately called a Hossfield pyramid, named for the late Casper Hossfield of the AAVSO Solar Division. This device is simply a pyramid-shaped box, constructed of lightweight materials. Thin wood or cardboard may be used to advantage in this design. The small end of the box is securely attached to the projecting eyepiece, with the viewing screen located at the base of the pyramid. The interior of the box should be flat black to prevent unwanted reflections, and be sure to leave one side of the pyramid open to allow viewing of the projection screen.

Care must be taken when selecting an eyepiece used for solar projection. The intense heat present at the focus of a telescope objective could possibly damage an eyepiece beyond repair. Optical cement used between the lens elements has melted in the past, forever ruining an eyepiece. Of course, if too large a telescope

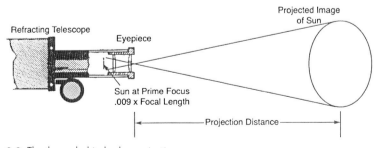

Figure 3.2. The theory behind solar projection.

is used for projection, heat might shatter the glass of the field lens of an eyepiece. To limit these situations, a maximum objective diameter of about 100 mm (4 in) is recommended for solar projection. An orthoscopic eyepiece, providing a better-quality projected image, has been used with a small aperture telescope (less than 80 mm), resulting in no damage to the eyepiece. For an observer with a larger telescope, either the classical Huygenian or Ramsden eyepiece is preferred. These are both non-cemented eyepiece designs, containing only two lens elements. Because the field of the Ramsden is curved, a Huygenian lends itself better for projection. Huygenian and Ramsden eyepieces are available at most low-end department stores that sell telescopes, and through a number of surplus optical dealers advertising on the Internet. Select only an eyepiece with a metal barrel and lens cell; plastic eyepiece parts may be susceptible to melting from the Sun's heat.

To begin observing via projection, an observer needs to decide upon a size for the projected solar image, and then find the eyepiece to screen the distance necessary to obtain that size. Simply holding a sheet of stiff white paper or cardboard some distance from the eyepiece and focusing with your free hand can give you an idea of the distance necessary. If the separation becomes so great as to be impractical, then switch to a shorter focal length eyepiece. Once the position giving the desired diameter of the Sun is found, make a note of the distance between the eyepiece and screen. Remember, the Sun does not have a fixed angular size in the sky. It will appear larger in December than in June. Unless provision is made to somehow vary the distance between the eyepiece and the projection screen throughout the year, the Sun will appear to grow and shrink in size. For casual observing or making sunspot counts, this is of no concern, but to the observer involved with heliographic position measurements, the Sun needs to be a uniform projection size year round (Figure 3.3).

Other than the trial and error method just discussed, the solar observer can find an approximate distance measurement for constructing the projection screen mathematically. A projected image of the Sun with a diameter of about 150 mm (6 in) is adequate for showing an interesting, detailed survey of the photosphere. With this size, the brightness of the projected Sun will likely be at a comfortable level, but note that with any given telescope, a smaller projected image is brighter and a larger image dimmer.

The first task is to calculate the diameter of the Sun as it appears at the prime focus of the telescope. A short focus telescope produces a small disc diameter, and a long focus instrument a larger. An approximate year-round diameter can be obtained by multiplying the focal length of the telescope by 0.009, using the same unit of measure throughout. As an example, let's select the typical 102 mm aperture telescope with a 1500 mm focal length. The Sun's virtual image at the prime focus of this instrument will be 13.5 mm (1500 mm × .009 = 13.5 mm) in diameter. If we wish to have a solar disc size of 150 mm on the projection screen, a magnification factor of 11.1× is necessary, found by dividing the desired projected size by the diameter of the virtual Sun (150/13.5 mm = 11.1).

Most likely you will have a particular eyepiece in mind for projection. One with a focal length of 12–28 mm is usually about right. A shorter focal length may not cover the whole disc of the Sun, and a longer focus eyepiece may require an excessive projection distance. For our example, an eyepiece of 25 mm focal length is selected. With the projection magnification (11.1×) computed, and knowing the focal length of the projecting eyepiece (25 mm), simply insert these

Figure 3.3. A Hossfield pyramid made of black foam board, white card stock, spray adhesive, and masking tape.

numbers into the following formula to solve the projection distance (separation of screen and eyepiece).

Projection Distance = (Magnification + 1) × Eyepiece Focal Length

Solving the equation tells us that separating the eyepiece and projection screen by 302.5 mm (12.3 in) will give our desired, 150 mm diameter projected image of the Sun.

Here are some additional tips for the solar projectionist. Make a vigorous effort when providing a shaded environment for the projection screen; the shaded screen facilitates increased contrast of the image. Always keep a watchful eye toward those who may inadvertently try to "sneak a peek" though the telescope. As with any solar observing, remove or cap the finder scope. Keep the viewing screen surface and field lens of the projection eyepiece clean and free of dust, or additional "sunspots" and "pores" may appear that are really non-existent. Part of the regular observing routine should include directing the telescope away from the Sun periodically for a slight cooling period. Finally, when the observing session is complete, let the telescope cool down before removing the projection eyepiece. You may be surprised to find the chromed barrel hot to the touch.

Direct Observation

Although projection is the least expensive and the safest method of observing the white light Sun, with this technique there is a tendency to miss the finest details. Direct viewing through the telescope permits a superior view and is the most popular method of observing with today's amateur astronomer.

Some early telescopic observers were brave but rather foolish experimenters. They would point a telescope sunward when it was low in the sky, or diminished by thin clouds. In those days, infrared light was unknown, as well as the damage it does to eyesight. Sadly, one of the premier scientists, Galileo, was blind by 1640. There are a few accounts of early astronomers attempting to dim the brightness of the Sun by using so-called "colored screens" within the telescope – a precursor of what was to follow.

However, very little changed concerning those poor observing habits until the late 1700s. William Herschel then took a slightly different direction, by using a 300 mm (12-inch) Newtonian equipped with a solar filter positioned between the eyepiece and diagonal mirror. This unique filter consisted of a "water tight" container with polished entrance and exit windows. Herschel filled this container with various colored liquids to filter, and dim, the brightness of the Sun through his telescope. This may have worked for him, but think of the heat present at the filter's location near the eyepiece!

Also in vogue during the 1800s was the use of "smoked glass," made by holding a piece of glass over a candle flame until a layer of soot was deposited. To seal the soot in, another piece of glass was placed over the first, and the two panes bound together with tape. For use, the observer would position the sandwich between the eye and the telescope's eyepiece, hopefully finding an area of sooty deposit sufficient to dim the Sun, all the while risking the glass shattering next to the eye! These were all unsafe attempts at directly viewing the Sun and should *never* be duplicated.

Fortunately, these futile efforts have been banished to history. Today, the solar astronomer has knowledge about the danger of solar observing and has at his or her disposal a variety of safe, commercially made observing appliances available for the telescope. Direct white light observation of the Sun can now be accomplished safely and confidently.

Objective Filters

The vast majority of amateur astronomers use an objective filter for white light solar observing. These filters are intended to screen out over 99.999% of the sunlight before it enters a telescope. An objective filter is always placed at the entrance to a telescope, effectively filtering all incoming light. The ability to discern fine solar detail, evaluate seeing conditions, and eliminate heat within the telescope when using an objective filter are distinct advantages for the observer.

Objective filters come in several flavors. That is, they have features that set them apart. The choice of substrate, or the material used to support the light-rejecting element of the filter, is one consideration. A significant portion of a filter's cost and performance is determined by the choice of substrate. Optical density, or how greatly sunlight is diminished, is another very important feature. Optical density is an expression of the reduction in light passing through a filter as a power of ten. For example, a filter of 5.0 density corresponds to a reduction of 10^5, or 100,000 times $(10\times10\times10\times10\times10)$.[1] Sometimes a filter is given a thin density coating, making it particularly suitable for photographic purposes but unsafe for visual applications. A visual observation filter usually has a density value of 5.0; a photographic-only version is many times in the 3.8–4.0 range. The color transmission characteristics of an objective filter determine which solar features the filter will enhance. Filters that transmit toward the red end of the spectrum favor sunspot detail; those toward the blue work well for observing faculae and granulation.

There are several forms of substrate available in the marketplace. The two widely used products are glass or a mylar/foil material. For the casual solar observer, neither really has a particularly strong advantage over the other; either type is suitable for occasional sunspot viewing or seeing the partial phases of a solar eclipse. For occasional observing, the glass filter might be preferred, but only because of its durability. Those of you with a specific interest in daily observations of the Sun might be more selective in your choice.

The glass objective filter has been available in the amateur astronomy marketplace for a number of years. During the 1960s, Optron Laboratory of Dayton, Ohio, offered a range of quality glass filters for owners of various telescope sizes. Bausch & Lomb marketed a series of safe full aperture solar filters through their Micro-Line division in the early 1970s. Today, several manufacturers of the glass-type solar filter exist to meet the needs of the amateur astronomer possessing a commercial or custom-made telescope. Refer to the list of equipment suppliers near the back of this book.

To make a glass solar filter, several thin layers of nickel, stainless steel, and chrome are vacuum deposited on the glass. Sometimes, alloys such as Inconel (nickel-chromium) are chosen as the filtering medium. Regardless of the material in use, the goal is to block the necessary amount of light across the spectrum, including UV and IR. A good policy for an amateur to follow is to request assurances from the manufacturer as to their product meeting that safety requirement. A glass filter is normally long lasting and quite durable.

Usually, the glass or mylar filter comes mounted in an attractive, polished aluminum cell that slips over the entrance to the telescope. A felt liner or a set of nylon screws guarantees snugness to the telescope tube. When making a purchase, it is necessary to know the exact diameter of the telescope tube or the objective lens cell a solar filter will be expected to fit over. Unless the make and model of the telescope is specified, always measure and order the filter cell a few millimeters larger than the diameter of the telescope tube. In some cases, unmounted filters can be purchased for adding to a custom-made holder. If you are handy in this way, there is nothing wrong with assembling your own solar filter.

The Sun viewed through a glass filter typically appears yellow-orange. This is an advantage for those with an interest in sunspot studies. A filter transmitting primarily in this region of the spectrum will boost the contrast and show much detail in the penumbra of a sunspot. Although sunspot detail may be improved with this filter type, other features such as faculae and granulation tend to be diminished. Expect to see only the brightest faculae with a yellowish-orange biased filter. Observers seeking a wide-scope of observational possibilities will consider obtaining a solar filter that transmits a wider spectral range. Filters that do this transmit a neutral- or white-appearing solar disc.

When an objective filter is placed within the light path of a telescope, it becomes part of the optical train of that instrument. Consequently, how it affects the final output, or wave front, of the light entering the eye is a critical factor in its selection. Said another way, the optical quality of a filter is determined by how little it distorts what is viewed through the telescope. A high-end glass solar filter is, indeed, a well-made optical flat, deposited with a metallic coating. Both sides of a filter must be parallel, or a condition called wedge occurs. Wedge causes the filter to act like a prism, creating color defects similar to atmospheric refraction. The two surfaces of a glass filter must also be flat and extremely smooth. In an ideal situation, the surfaces will be at least equal in quality to the remaining optical components of the telescope.

Think of the performance of a telescope as being only as good as its poorest made component. When seeing effects are taken into account, a filter-created distortion is not always noticeable to the eye, particularly at the low magnification used for whole disc viewing. Poor seeing conditions regularly mask a poor-quality solar filter. It's when magnification is increased, as when attempting to see fine detail in a sunspot when the atmosphere becomes steady, that a poorly made filter becomes obvious. Granulation will not be apparent, pores may wash out, and fine penumbral filaments that should be visible are found to be beyond the resolution of the telescope. Such are the results of observing with a poorly made objective filter.

Since an optically flat plate is difficult to manufacture, it is expensive to produce, and the cost of a proportionately larger size increases dramatically. A smaller than full aperture (as in off-axis) size costing less may be obtained, but at the price of reduced resolution. There are inexpensive glass filters manufactured that use plate, or what is called float, glass. Often these are not worked to optical flat standards and may disappoint a serious solar observer when high-resolution studies are desired. Because of the time and expense involved in producing a well-made optically flat glass, the cost of such a filter will usually be an indicator of its quality.

In the early 1970s, amateur astronomer Roger Tuthill experimented with the use of a new material for the purpose of constructing a solar filter. His product, Solar Skreen, uses two layers of aluminized, optical grade Dupont mylar, each with a thickness many times less than that of the human hair. These filters are so thin that degrading optical effects are kept to a minimum, allowing quality observations to be made. A Solar Skreen filter transmits largely toward the blue end of the spectrum, giving the Sun a pale bluish-white cast. Although some observers find this to be aesthetically displeasing, it does enhance the contrast of faculae and granulation noticeably. The blue cast also makes a white light flare easier to spot. On the down side, a Solar Skreen filter contributes to light scattering in the telescope, which is further reinforced by the atmospheric scattering of blue light.

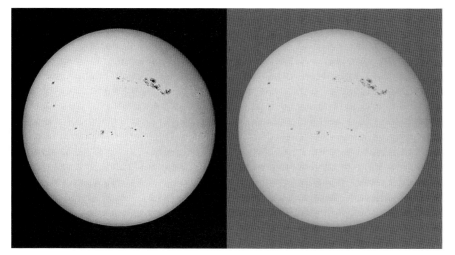

Figure 3.4. The effects of excessive light scattering within the telescope are apparent in these two images. The photo on the left exhibits minimal scattering, with a dark background sky and contrasting detail on the solar disc. The Sun on the right is seen through a filter contributing to light scattering. Jamey Jenkins.

Viewed through a telescope fitted with one of these filters the background sky up to the edge of the Sun may appear brighter than expected. This background scattering effect can be improved with a Wratten #21 orange filter at the eyepiece used with the Solar Skreen filter at the telescope entrance (Figure 3.4).

Like glass, a mylar-type solar filter is available from several manufacturers. Specific standards in the manufacturing of the mylar must be met, or poor viewing will result. The mylar must be of optical grade, possessing a uniformity in density and thickness, with an aluminum coating that blocks sunlight from the ultraviolet to the infrared. The cost of a mylar filter is less than an equivalent diameter glass filter, opening up direct viewing possibilities to a number of budget-minded observers.

During the late 1990s, a new filter material found its way to the marketplace from Baader Planetarium in Germany. Marketed as Baader AstroSolar™ Safety Film, this product has demonstrated amazing optical performance, with transmission characteristics suitable for a variety of solar observing projects. Although not truly a mylar – it's known as a foil – it is similar in appearance to mylar. The substrate performs nearly as well as a ground and polished optical window, while the cost savings of the foil filter compared to a high-quality glass is truly, astronomical. The Baader solar filter is a single layer film, coated on both sides, transmitting light in a neutral fashion; that is, the Sun appears white through it. Matching a colored or narrow band eyepiece filter with a Baader filter boosts the contrast of selected white light features. Additionally, when compared to other mylar filters, the scattering of light in a telescope is lessened markedly as seen through the foil product.

Baader AstroSolar™ Safety Film is without a doubt one of the most successful new products on the solar observing scene. Being a thin foil material, it is still susceptible to damage from careless handling, even though the substrate and coating are durable and tougher than other aluminized mylars. All things considered, a Baader solar filter is perhaps the best direct viewing objective filter available for an amateur solar astronomer.

As you can see, several factors should be considered when selecting an objective solar filter. These factors range from which feature you choose to observe to the funds on hand. The rule of thumb when making a filter purchase is this: Secure the finest quality filter you can afford. It's pointless to obtain a poorly made objective filter that distorts the view through a first class telescope. Discussing with other amateurs any filter they have used in the past, or are currently using, is perhaps the best way to become knowledgeable. The ability to borrow a filter for a time, and "try it out" on your own telescope, provides immediate insight into a product's usefulness. Read the reviews posted by other amateurs at various websites and on message boards. Then make your choice, knowing what to expect from the objective filter you've selected (Figure 3.5).

After obtaining a new filter, perform a few tests before directing it toward the Sun. Begin by holding the filter between your eyes and a 100–150 W light bulb. Look for any pinholes or defects in the filter's coating. If present, these imperfections will allow unfiltered sunlight to enter the telescope, resulting in possible eye damage. Don't use any filter that exhibits excessive pinholes, scratches, or visible regions having unequal density to the coating. Several tiny pinholes can be painted over with an opaque Sharpie marker or black paint on the inner side of the filter with no harm done. Anything greater than a few minor imperfections would indicate a poor coating job, particularly on a new filter.

If the objective filter appears fine, attach it to the telescope's entrance. A filter must be mounted so that it can't be accidentally knocked off the end of the telescope. If any opportunity exists for an accident, masking tape can be used to secure the filter in place. Once the filter is attached, point the telescope toward the Sun, using either the telescope's shadow on the ground or a sunfinder device as a guide.

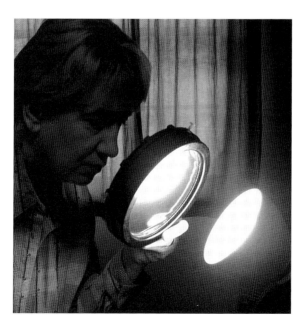

Figure 3.5. Check for defects such as pinholes, scratches, or a nonuniform coating on a solar filter before attaching it to a telescope.

The following additional tests should be performed on a day with good seeing conditions and good to excellent transparency. Using a low magnification eyepiece that shows the entire solar disc, look for the amount of haze surrounding the Sun. If the sky is blue, with little bloom showing when your outstretched fingers block the Sun, but a noticeable amount of haze is surrounding the solar disc through the telescope, it's likely the filter is scattering an excess of light. This may be due to a defect in the filter or it could be just an inherent quality of the filter, particularly a mylar type. Now, change eyepieces to one of a higher magnification and center a small umbral spot in the field of view. By slowly moving from a point of inside focus to a point outside, note whether the spot appears to change shape. It should focus symmetrically, that is, maintain its roundness. If it becomes elongated on one side of focus and shifts elongation perpendicularly on the other side of focus, this indicates astigmatism. This defect will be visible only with a glass filter and indicates that either the filter is of poor optical quality or maybe it is pinched in its cell. This same test should be performed with the same telescope and eyepiece combination one evening on a star, without the filter in place, to confirm whether the telescope or the solar filter is at fault. If the filter is the culprit, change the filter.

Objective filters should be stored in a dry, dust free box or container when not in use. You might use a corrugated cardboard box, but you could also use a plastic ware container with a snap-on lid successfully. If cleaning a filter becomes necessary, and eventually it will, start the process by gently brushing away dust or dirt with a soft cotton ball, blower brush, or by using a can of compressed air. If this doesn't do the job, a glass filter may be *gently* cleaned using lens cleaning solution and a soft cloth or cotton ball. Baader AstroSolar™ Safety Film may be similarly cleaned with a mild detergent and distilled water. Always be careful of damaging the coating, and inspect the filter for pinholes and scratches before putting it back into service. Mylar-type filters are best left to cleaning with a soft brush or compressed air only; the coatings are particularly fragile.

Several makers of objective filters supply a product with a thin rejection coating that increases the light throughput of the filter. For direct visual observation these filters are unsafe, but for photographic use, they are ideal. The increase in light transmission permits a photographer to draw upon a higher shutter speed in his or her camera, effectively "freezing" turbulent seeing conditions and limiting the atmospheric smear in a photo. The abundance of light from a photographic version filter also makes it practical to include a Wratten or interference-type filter to the filtration pack. For photography with a thin density objective filter, an electronic camera equipped with an off-camera video monitor is the safest and preferred method.

Herschel Wedge

Before the advent of the objective filter, many solar observers found the use of a low-angled prism to be an important addition to their collection of accessories. Also referred to as a sun diagonal, the common name for this device is a Herschel wedge. In the early 1800s John Herschel, William's son, advanced the method of using a thin wedge or prism-shaped piece of glass to reflect about 5% of the incoming light of the Sun to a telescope's eyepiece. A secondary filter to further dim the light to a safe level was placed between the prism and the eyepiece. This technique worked because the only light reflected to the eyepiece was from the front surface of the prism. A small portion of the heat and light was absorbed within the prism, and the remaining 90–95% passed through and out its back side. Secondary reflections into the eyepiece did not exist, because the angle of the prism (10°) prevented ghosting from occurring.

The Newtonian reflecting telescope has been adapted to use a Herschel wedge as the diagonal mirror, but with some annoying difficulties. Since 95% of the light and heat of the Sun is discharged out the rear of the prism, undesirable reflections in the tube and pockets of heated air are created within a Newtonian. A catadioptric is out of the question for the same reason that solar projection is not suitable – possible damage to the telescope from internal heat. The refracting telescope, with its straight-through design, is ideally suited for using a Herschel wedge (Figure 3.6).

Several firms manufacture a prism-viewing system for use on your personal telescope, Baader Planetarium and Intes being two well-known distributors. Both of these products are available with a filter kit that includes suitable neutral density (ND) filters that MUST be used with the Herschel wedge when solar observing. Since the prism is housed in a star diagonal, and set at what is called the "Brewster or polarization angle," image brightness may be adjusted with the addition of a single polarizing filter located between the eyepiece and wedge. Some amateurs choose to additionally insert an IR rejection filter as a safety buffer in this package. Again, always follow the manufacturer's specifications on the use of these accessories.

The white light view of the Sun through a Herschel wedge is untainted color-wise. The Sun appears crisp and contrasty, with a black background sky on days of little haze. Some of the products just mentioned can be used with refractors up to 180 mm (7-inches) aperture for extended periods, with no damage to the prism

Figure 3.6. Herschel wedge by Baader Planetarium. The supplementary neutral density filters are supplied to control excess brightness of the view through the prism assembly. Eric Roel.

from heat. However, we would recommend a regular cool-down period be instituted into the observing routine, especially if a daylong observing marathon is contemplated. This could be accomplished by pointing the telescope away from the Sun for 5–10 min every so often.

Because 95% of the energy from the Sun is exhausted through the back side of a Herschel wedge, caution must be taken to prevent eyes, fingers, etc., from being placed within these rays. The commercial units mentioned use an effective light trap system to dissipate heat and light, which prevents accidental burning or injury. Some homemade diagonals are enclosed in a metal box, with ventilation holes near the rear. These boxes could become hot to the touch, so be wary of any design lacking sufficient venting.

Using Supplementary Filters

Solar observers regularly use a colored glass eyepiece filter in combination with a safe direct view filtration to enhance the contrast and visibility of a white light feature. Earlier we mentioned that a glass objective filter transmitting primarily in the yellow/orange region of the spectrum is suitable for sunspot studies, and that a mylar filter biased toward blue works well for faculae and granulation viewing. The addition of a colored glass filter to either of these reinforces this effect by further narrowing the bandwidth of light transmitted to the eye. Remember this: **No eyepiece filter is ever to be used alone for solar observation; it should only be used to supplement a safe solar filter already in use.**

Glass eyepiece filters come in a seemingly infinite variety of hues and shades. The standard labeling system for an eyepiece filter is the Wratten number. Devised in the early twentieth century by Frederick Wratten, a photographic specialist from London, the Wratten filter was initially used to allow the photography of a specific color of light with the newly invented panchromatic film. The identifying system for a Wratten filter is universal, in that the characteristics of a same numbered filter will be nearly identical between makers. Some commonly available filters are the #11 (yellow-green), #21 (orange), #25 (red), and #56 (light green). Most quality eyepiece manufacturers supply filters for their products. Two sizes of eyepieces and filters are standard, the 1 1/4 in and the 2 in. Each size fits an appropriately constructed focuser. If you use the larger 2 in barrel, then the standard 48 mm camera filter can be substituted as an eyepiece filter. In nearly all cases the colored glass filter screws into the bottom of the eyepiece barrel.

Absorption is the method by which a simple glass filter works. Light passing through the glass has some wavelengths absorbed by aggregates, while other wavelengths pass right on through. Besides reducing the total amount of light transmitted, usually the denser or darker a color filter's shade is, the tighter is the "cut-off," or transmission, of the filter around a particular peak wavelength. The amount of light passed on either side of the peak transmission wavelength is known as that filter's bandwidth. An absorption filter is a broadband device that typically passes several hundred or more angstroms (1 Å = 0.1 nm) of light. An absorption filter is not intended for, or able to show, monochromatic features, such as prominences or flares; what it can do is enhance the visibility of relatively low-contrast white light features.

For purely visual studies, a lighter shade is the most useful. Denser shades, while more effective, particularly for photography, may create too dark a view visually,

Table 3.1. Color filter usage for solar observing

Color	Wratten filter	Application
Dark Red	29	Red and orange filters increase the contrast of knots and radial
Red	25A	streaks in the penumbrae of sunspots.
Light Red	23A	
Orange	21	
Yellow	11	Yellow...neutral, all-around filter that lessens achromatic color errors.
Light Green	56	Green filters increase the visibility of granulation and faculae.
Green	58	
Dark Green	61	
Blue	47	Blue will assist in seeing faculae further from the limb than normal.

making observation difficult. Table 3.1 lists Wratten filters ordinarily used for white light observing, and how they affect the Sun's appearance. Although some filters will work especially well when coupled with a red or blue biased filter, all are suitable if matched with a primary filtering appliance that transmits a neutral- or white-appearing Sun.

A red eyepiece filter, such as the Wratten #25, will darken a sunspot umbra, causing it to stand out against the solar disc. For an observer interested in the statistical counting of sunspots, this is a helpful tool for locating weak detached portions of umbrae, an umbra in the early developmental stage, or pores. A red filter is also known to increase contrast within a sunspot's penumbra. Streaks or knots appearing within the filamentary structure of the penumbra are amplified, thereby making those features more apparent.

The view of other solar features, such as faculae and photospheric granulation, is improved by using a green transmitting filter. A #56 or #58 will lighten a facula near the limb while darkening the surrounding photosphere. Granules will be improved but still remain difficult to spot until superior seeing conditions are encountered. Unless a facula is unusually strong, this feature is hardly seen deep into the solar disc. Most faculae are visible near the limb of the Sun, within the sunspot zones. A deep blue filter, particularly the Wratten #47, is helpful in viewing a facula nearer the center of the solar disc. A #47 combined with a visually safe solar appliance creates a deep violet Sun. As adults reach an older age, our eyes become less sensitive to this part of the spectrum, increasing the difficulty we experience in seeing violet hued light. Don't be surprised if you find you can photograph some solar features in this region of the spectrum easier than you can see them visually.

The use of a broadband filter permits the observer an increased opportunity for achieving telescopic resolution approaching 1 arc second. Recall that atmospheric refraction becomes greater than 1arc second when you observe over 25° from the zenith. A broadband filter, when inserted in the telescope, removes any additional colors resulting from atmospheric refraction. A narrow band filter is even more effective when used for this task.

Whereas a broadband filter transmits a wide portion of the solar spectrum, usually exceeding several tens of nanometers (nm), a narrow band filter passes only the thinnest slice of light from the spectrum. The narrow band filter is most often used for monochromatic observing; monochromatic means, literally, one color. Capable of showing prominences and chromospheric layer activity, a narrow band filter is necessarily very selective in the light it passes. Based on the interference of light, this filter transmits light in a region where a monochromatic feature is the brightest, where it's in emission. The bandwidth of a narrow band filter may be .1 nm wide, and often substantially less.

Between these two extremes, broadband and narrow band, is a class of filter operating on the same principle of interference as a narrow band filter but passing a bandwidth of light less than 10 nm and greater than 1 nm. Although not selective enough to allow extensive chromospheric observing, this class is particularly effective in the contrast enhancement of photospheric features, and to a limited extent, observing Calcium-K activity directly above the photosphere.

Baader Planetarium distributes what is marketed as a solar continuum filter, and also a Calcium K-Line filter. Continuum is a term used to define the combination of all the colors emitted by an object. In the case of our Sun, it is analogous to a white light view. Baader's solar continuum filter passes a 10 nm band of light near the 540 nm wavelength in the green portion of the solar spectrum. Coronado distributes a similar Fe XIV eyepiece filter having a bandpass centered near 530 nm. These filters are effective due to an emission near this wavelength that is linked to the solar faculae. Bandpass filters at various wavelengths, and with an assortment of bandwidths, are also available from many optical filter manufacturers. A green filter with a bandpass centered near 520—540 nm and a bandwidth of 10 nm or less works especially well for the features previously mentioned. The difference between a commercially available eyepiece filter and a filter obtained directly from the optical manufacturer will be a lack of threads on the generic filter to attach it to an eyepiece. Often an old screw-in type filter housing can be salvaged for use with a bandpass filter purchased directly from a manufacturer.

The Baader Calcium K-Line has a relatively wide 8 nm bandwidth filter centered on 395 nm to increase contrast of notable Ca-K features. This particular filter is recommended for photographic use only because of a potential for high UV exposure to an observer's eyes; in practice, it's mated with the Baader low-density (photographic version) white light objective filter.

Another interesting filter would be one passing light from what is called the G-band, which is in the blue portion of the spectrum. At about 430.5 nm is the G-band grouping of spectral lines, which go into emission during flare activity. Used photographically, or with other video observing means, a 10 nm or less bandwidth filter centered near the G-band will improve an observer's chance of locating solar flares in the solar continuum. These white light flares, as discussed later in this book, are relatively rare events.

Dedicated Telescopes

A telescope created specifically for observing the Sun in white light has features incorporated into its design that limit usefulness in other areas of astronomy. Although this may seem a hindrance to the average amateur astronomer, such a telescope provides superior performance when solar observing.

Commercially, there are a few dedicated telescopes on the market intended for white light observing. Those that are available set themselves apart from the standard telescope line by coating the primary lens (these are refractors) with the metal alloys used to manufacture a glass objective filter. Turning the objective lens into a solar filter is one way of assuring safety; the filter can never be accidentally knocked off one of these telescopes while observing. Additionally a benefit is gained by removing the external glass or mylar substrate, since one possible element distorting the Sun is eliminated from the light path.

A dedicated telescope is indeed handy for a quick look at solar activity, the viewing of partial phases of a solar eclipse, or for the observer interested in statistical studies, such as sunspot counting. One limiting characteristic of the available commercial instruments is aperture, and, in effect, resolution. These telescopes typically have a 102 mm or less diameter objective lens, borderline for the desired 1 arc second resolution. Secondly, the light rejecting coating when applied to the objective is given a density of 5.0, safe for visual studies but not necessarily desirable if an observer later becomes interested in photography. But for casual solar observations, these are excellent telescopes, making a wonderful addition to any amateur's collection.

If you enjoy constructing your own equipment, several designs are popular. We will highlight a few here to illustrate the possibilities and perhaps pique an interest for you.

Dobsonian Solar Telescope

In the 1960s John Dobson, of the San Francisco Sidewalk Astronomers, devised a variation on the Newtonian reflector that incorporated a unique observing safety feature not found on other telescopes. The instrument became known as the Dobsonian solar telescope, enjoying favor with telescope-making enthusiasts over the years. For the most part, this instrument is intended for a low-power view of the white light Sun, suitable for sunspot counting or just exploring the changing appearance of the solar disc.

The standard Dobsonian solar telescope (DST) uses a plate glass one-way mirror (available at mirror dealers) as a partially aluminized objective filter at the entrance to the telescope. Amateurs desiring a higher quality view may wish to substitute a polished glass that has been partially aluminized commercially to the necessary transmission level (5%), since the amount of wavefront error induced by typical plate glass is going to be left to chance. This front plate is positioned at a 45° angle to the unaluminized primary mirror. When the aluminized side of the plate is facing the primary mirror, the rear side of the plate serves as a Newtonian diagonal, directing light to the side of the telescope tube. This is where the unique safety feature comes into play. Should the front plate somehow become dislodged or broken, the telescope is effectively shut down; it becomes inoperable. Other standard Newtonian telescopes may continue to emit light if the normal objective filter is displaced or broken. The DST is the only solar telescope with this feature.

The subdued light of the Sun transmitted through the entrance plate is again reduced by the 4–5% reflecting of the uncoated primary mirror. In front of the eyepiece in a DST is an appropriate shade of welder's glass, which dims the visible light and removes unwanted IR and UV light. This is the only instance where welder's glass may be used in a telescope for solar observing. *Never* insert a welder's glass (even the #14 shade) in front of the eyepiece of a telescope not having a prior means of heat and light reduction, because the filter will crack! The choice of shade for the welder's glass – they are available from #14 (darkest) to #2 (lightest) – will depend on the transmission characteristics of the front one-way mirror. Some experimentation on the part of the observer may be necessary to select a correct shade of filter. For a one-way mirror transmitting approximately 5% of the incoming light, a welder's glass shade of 7 or 8 is about right, in a typical DST (Figure 3.7).

What is a typical DST? Often the telescope is constructed with a 150 mm (6-inch) aperture and a focal ratio of f/10. A larger aperture is often limited by atmospheric seeing conditions. The long focal length provides clean, sharp solar images. An f/10 primary mirror can retain an elementary spherical shape, making the construction for an individual grinding his or her own mirror a simple task. The heavy front one-way mirror plate of a DST requires creative balancing of the tube with weight added near the primary mirror's end. To reduce light scatter within the telescope, baffling around the primary mirror mount is recommended, and if the tube is a bit larger than the primary mirror, any internal air current detrimental to seeing conditions will be minimized. Ordinarily, a DST is mounted on a wooden alt-azimuth rocker box. This arrangement permits convenient to and fro movement, while requiring only minimal space when the telescope is stored.

Having myself viewed the Sun through several of these telescopes, I can testify that the better-made examples are comparable to the view through a typical Newtonian with the standard white light objective filter in place. For the telescope maker wishing a dedicated instrument intended for a low-power white light view

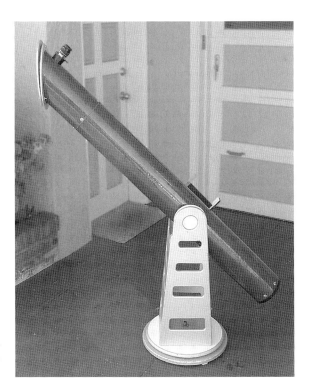

Figure 3.7. A Dobsonian solar telescope with a 150 mm f/10 primary mirror. Ray Cash.

Figure 3.8. Art Whipple's 203 mm photographic Newtonian. The inset is an image of AR0908 obtained on September 10, 2006, under seeing conditions of 1 arc second. Art Whipple.

of the Sun, a Dobsonian solar telescope is ideal. Safety and convenience earmark the fine points of a novel design.

Photographic Newtonian

For an amateur desiring a telescope capable of high-resolution imaging of the Sun, this custom telescope by veteran solar photographer Art Whipple is enlightening. A strictly photographic model, the telescope is built around an unaluminized 203 mm (8-inch) f/10, full thickness pyrex paraboloidal mirror. As in a Dobsonian, the primary mirror reflects 5% of the incoming light of the Sun, a substantial reduction in brightness but not sufficient for direct visual observation.

Taking advantage of the fact that a white surface reflects heat, Whipple has constructed a Newtonian telescope optimized to provide as little disturbance to the seeing conditions as possible. The mirror mount is painted white in order to minimize heat buildup that would otherwise distort the figure of the primary optic. The open truss tube eliminates any internal air currents that might disturb the wavefront. Although light scattering is greater with this design, the flat black interior surfaces keep it to a minimum, and image contrast is at an acceptable level. A standard coated Newtonian diagonal mirror is used to eliminate heat buildup in the optical train.

No viewing eyepiece is available on this telescope; it is skillfully built to be completely electronic. The remote focusing unit contains a 2.3 mm diameter field stop, or diaphragm, to prevent any unwanted light from entering the imaging optics, another deterrent to scattered light. Enlargement of the primary image to a scale allowing diffraction-limited performance is accomplished with a 16 mm focal length projection lens, borrowed from a microfilm printer. Before reaching the detector, a Pulnix TM-72EX B/W CCD video camera, the light of the Sun passes through a 9 nm bandwidth interference filter centered on 520 nm.

Slewing, focusing, image acquisition, and monitoring the field of view are done from an enclosed control room located nearby. Being able to control the telescope from a separate location provides a measure of convenience while removing the observer – yet another source of thermal disturbance from the immediate vicinity of the instrument.

The photograph shown here of the Whipple telescope illustrates the care that has been taken in its construction. Note the light and heat protective blanket covering the imaging unit, an attempt to control local thermal effects. Solar observers, to be successful at imaging, must make a conscious effort to optimize equipment and improve local seeing conditions. The results obtained with this telescope are a testimony to such efforts.

A Final Word

Whether you adapt a conventional telescope, purchase a specific model intended for white light observing, or construct a dedicated solar telescope, the key to success in solar observing is found in relatively simple, well-made, medium-aperture optics.

It's in solar observing that the effects of image aberrations, scattered light, and the heating of optical and mechanical components become extremely damaging to the output of the telescope. Because many solar features are of low relative contrast, such negative effects tend to wash out what is seen through a telescope. The delicate details of the Sun require an optic capable of delivering near its theoretical resolution. The more components existing within an optical system, the greater are the opportunities to introduce scattered light, destructive aberrations, and so on. This is one reason why a refractor is the preferred telescope for the solar observer; the instrument issues are easier to control.

So, the ultimate goal with any telescope intended for serious study of the Sun is to keep the telescope simple and optimized to deliver sharp and highly contrasted images. Any telescope design that follows those criteria will enable the solar observer to enjoy wonderful views of activity on our nearest star, the Sun.

Reference

1. *Solar Observing Techniques*, C. Kitchin, Springer-Verlag, 2002

Chapter 4

White Light Solar Features

Bubbling Witch's Brew

If any statement ever rang true regarding solar observing, it is this one: the only thing unchanging about the Sun is that it's ever changing. Sound confusing? It shouldn't. Simply put, our Sun is in a constant state of flux. Even during the period of minimum solar activity, the Sun churns like a caldron of bubbling witch's brew. Granules are forever developing and dissolving. Pores form, sometimes growing into spectacular sunspots, only to decay back to nothingness. The fact is, on any given day something spectacular and new can and often does occur on the Sun. Here again, we discover why amateur astronomers observe the Sun; each observation is unique and is never exactly repeated.

The surface of the Sun is the layer we call the photosphere, which literally means "sphere of light." Sunspots are the most obvious features of the white light Sun. They are dark patches that develop within a specific zone of latitude, changing position from day to day as the Sun rotates on its axis. Watching the daily development of a sunspot is one of the more interesting activities of white light observing. Sometimes within a sunspot a bright feature, called a light bridge, will grow, dividing the spot into parts. Mistaken occasionally for a transient event called a white light flare, a large light bridge often signals the beginning of the end for the sunspot. A pale, wispy, "cloudlike" feature called facula is sometimes seen surrounding a sunspot group as it emerges from the east limb, or as it makes its way around the back side of the Sun.

White light observing of the Sun doesn't require expensive, sophisticated equipment. A safe observing station can be readied with the addition of only a simple projection screen or a white light objective filter on your telescope. What will be in demand is attention of the observer to detail, because many solar features are of low contrast and are finely formed structures. Patience with daytime seeing conditions is essential, and knowledge of what the Sun has to offer is indispensable in determining what is visible.

Directions on the Sun

Before an explorer begins a journey, it's always advisable to gain an understanding of the layout of the territory and local directions. Which way *is* north, south, east and west? The same applies for any exploration of the Sun. To share your adventures with other explorers, one of the reasons to explore, it is necessary to establish where you were, in addition to what you've seen.

J.L. Jenkins, *The Sun and How to Observe It*, DOI 10.1007/978-0-387-09498-4_4,
© Springer Science+Business Media, LLC 2009

Compass directions on Earth are fairly easy to understand. If one faces toward the north celestial pole, the northern hemisphere point around which the heavens appear to rotate, south is directly behind you with east to your right, and west to your left. On Earth, the Sun, due to the rotation of Earth, appears to rise from the eastern horizon and set in the west.

Defining directions on the Sun is not so difficult either, if you picture the disc of the Sun as it appears in the sky. That is to say, the northern half of the Sun will be toward the north celestial pole; the solar southern region consequently faces the south celestial pole. The east-west limbs of the Sun are defined in a similar manner; the east limb of the Sun faces the eastern horizon of Earth, and the west limb faces the western horizon.

Through a stationary telescope, the western limb of the Sun will drift out of your field of view first, and, as the Sun rotates, features disappear behind the western limb. Conversely, the eastern limb of the Sun will drift out of the telescopic field last, and solar features first appear from behind the eastern limb.

North and south on the Sun can be determined through a telescope by gently nudging the instrument either north or south and noting the Sun's movement. For example, when you tap the telescope toward the north, the southern hemisphere will start to leave your field of view.

Understand that these are celestial directions. Due to the inclination of the Sun to the ecliptic and the tilt in Earth's axis, the Sun as viewed from Earth appears to tip and nod in the sky throughout the year. Depending on the date an observation is made, the Sun's true north or south pole, the axis around which it rotates, could be pointing toward or away from us by up to $7°15'$, and slanted to the east or west by as much as $26°21'$.

Some observations require no more than knowing the rough locations of N–S–E–W on the Sun, but for the earnest observer, precise directions are necessary. Accurate orientation of the Sun will be discussed further in this chapter later, when heliographic coordinates are the topic.

Active Regions

When an area of the Sun contains a confined, temporary event such as a sunspot, plage, facula, or flare it is known as an active region (AR). All active regions form because of the strong magnetic influence found there. Much of what solar astronomers observe in the photosphere is associated with active regions.

In order to maintain a record-keeping system of solar activity, astronomers devised a numbering plan that went into effect on January 5, 1972. Since that time, as each area of activity has been detected, that active region has been assigned a sequential four-digit number (i.e. 2054, 2055, 2056...). The National Oceanic and Atmospheric Administration (NOAA) is delegated the responsibility of assigning an "AR" number to each new event. A typical name for a region might be AR6092. Sometimes a solar event can last for several rotations of the Sun; in that case a region is given a different AR number for each appearance. Because AR numbers are limited to four digits a logical question to ask is, "What will happen when active region number 9999 is surpassed?" That did happen on June 14, 2002, when AR10000 was observed. The solution was that the four-digit numbering sequence was retained; the new fifth digit is just ignored. For example, AR10165 is referred to as AR0165.

Using resources found on the Internet, the amateur solar observer is able to determine the assigned number for any active region. Several sites exist that provide a daily Sun image with labels superimposed identifying the currently visible ARs. One frequently visited site is the web page of Mees Solar Observatory, University of Hawaii, which provides a daily active region map as well as whole disc white light and monochromatic images; their archive of earlier images and maps is helpful when doing research into past activity centers.

Rotation of the Sun

As discussed earlier in this book, the Sun is not a solid body but is composed of gaseous plasma. Because of this characteristic the outer layers rotate differentially, as though they were made of liquid. This results in the equatorial zone of the Sun completing a rotation before the areas near the poles. The Sun rotates as seen from above its north pole in a counterclockwise direction, east to west.

Solar rotation periods are of two types, the synodic and the sidereal. The synodic rotation period is the apparent rotation of the Sun, as seen from Earth. This is not an accurate or true rotation period, because our point of reference (Earth) continues in its orbit as the Sun completes a rotation. A sidereal rotation period is the time required for a position on the Sun to complete one rotation, as seen from a fixed point in space.

A sunspot, used as a marker, is seen to complete a sidereal equatorial rotation in 25.38 days. A spot $30°$ on either side of the equator takes about 27 days and, if located in the polar region, in excess of 30 days. The mean synodic period is 27.28 days, varying throughout the year due to the eccentricity found in Earth's orbit.

Astronomers define a time period on the Sun by its rotation. This is how the previous naming example of AR10165 is differentiated from the original AR0165, by knowing in which rotation period the event was visible. Rotation periods are based on Richard Carrington's observations from the 1850s, at Greenwich Observatory. The Carrington rotation number identifies the solar rotation as a mean period of 27.28 days, each new rotation beginning when $0°$ of solar longitude crosses the central meridian of the Sun as seen from Earth. Carrington, an English astronomer, spent years studying the accurate positions of sunspots, and from this information he computed the precise rotation rate for various latitudes of the Sun. Information regarding the start and duration of a Carrington rotation can be found on the Internet, in the Ephemeris at the back of this book, or in most of the many astronomical almanacs published yearly.

The Solar Cycle

In 1826, Samuel Heinrich Schwabe, a German pharmacist, began watching the Sun in a project designed to discover a planet inside the orbit of Mercury. The sunspots he observed by way of elimination were only a mild curiosity. His search continued methodically for a number of years, as the pharmacist failed to discover an inter-Mercurial planet, but what he did discover took all by surprise. From his routine searches, Schwabe recognized a periodicity to the number of visible sunspots.

A definite waning and waxing of solar activity became apparent. He announced his discovery as a 10-year recurrence pattern, and further observations confirmed his conclusion.

After many decades of continued study, the 11-year sunspot or solar cycle is a well-established fact. It's called an 11-year cycle, but that's really just an approximation; past cycles have had a period ranging from 8 to 14 years, the statistical average being about 11.1 years. Solar cycles are also given a number, just like active regions and rotations of the Sun. Astronomers dubbed cycle number one as commencing during the year 1755 and ending about 1766. Subsequent solar cycles have been well documented, providing a valuable database representing past histories of solar activity.

At the beginning of a cycle solar activity, for the most part, is in a lull; this time is called the solar minimum. The inverse is true for what is called the solar maximum, when sunspots and other activity are relatively abundant. The rise and fall during a cycle is not balanced. On average, activity following the solar minimum takes about 4.8 years to peak and another 6.2 years to decline before minimum is again attained. Solar astronomers try to predict the duration and strength of each new solar cycle, but an accurate forecast is difficult to obtain. This would indicate that mechanisms we do not yet understand are at play in the Sun (Figure 4.1).

A solar cycle begins with sunspots appearing in the high solar latitudes, away from the equator. As a cycle ages, and the numbers of daily sunspots increase, new spots begin to appear closer to the equatorial region. When a cycle ends, new spots begin appearing again in the high latitudes, signaling the start of a new solar cycle.

A sunspot is a magnetic feature and has a polarity of positive and negative. In either the northern or southern hemisphere of the Sun during a given solar cycle, the polarity of sunspots are uniformly arranged, with the leading sunspot in all bipolar groups having a positive polarity; the following spot in each group is negative. Interestingly, a sunspot during the same solar cycle in the opposite hemisphere has a reverse polarity of those in the other hemisphere; that is, the leading spot is negative and the following positive. An even more interesting phenomenon takes place when a new solar cycle begins. A reversal of the magnetic field happens, so that each hemisphere's spots have a polarity opposite of that from the previous cycle. Some astronomers use this field reversal as the gauge to judge when a new solar cycle has begun.

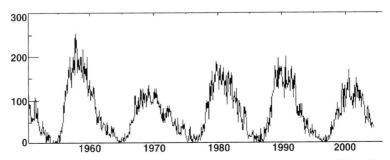

Figure 4.1. Solar cycles of approximately 11-years' duration are plotted prior to 1960 until near the end of Cycle 22 in the year 2005. Courtesy of NASA.

In order for the magnetic polarities to return to their original state, two solar cycles will have had to occur. This reversal and return of the Sun's magnetic field to its original state is called the magnetic cycle and averages 22 years.

Limb Darkening

The disc of the Sun suspended against the background sky has a three-dimensional effect in which the Sun appears brighter at the center and becomes, in a radial fashion, dimmer as you approach the limb. This gradient effect is called limb darkening and to some extent is the result of increasing temperature, as we look toward the core of the Sun. The outer layers are cooler and therefore darker-appearing to the eye. It is also important to understand that when an observer looks near the edge of the Sun, he or she is not observing as deeply into the solar interior as at the center. The angle involved when observing the limb causes our perceptive depth to be diminished, due to the cumulative opacity of the gas nearer the limb. Because of this opacity astronomers see slightly deeper into the Sun at the center, where the temperatures are greater and the appearance brighter.

Across the spectrum, limb darkening is not a uniform effect. When observed in the infrared (IR), it is practically invisible. Nearer to ultraviolet light (UV), the darkening is pronounced and more evident. At extreme ultraviolet wavelengths the limb is brighter than the center of the Sun, an effect called limb brightening; this, however, is beyond the observing ability of an amateur astronomer (Figure 4.2).

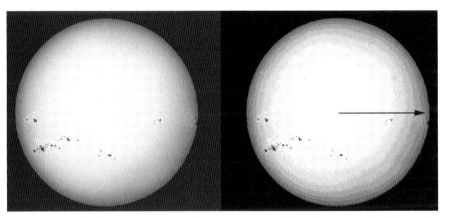

Figure 4.2. Limb darkening is apparent in this image courtesy of Gordon Garcia. The photo on the *right* has been digitally processed, using isophote contours to illustrate the degree of gradation from the *center* of the Sun outward.

Granulation

Over the entire photospheric surface is a textured pattern called granulation. Granulation is a derivation of the Latin word "granulum," implying grain. Observers seeing the granulation pattern have likened it to the appearance of rice grains, kernels of corn, or other geometric figures, including multi-sided polygons or elongated structures. Regardless of their shape, which is varied, the individual components of granulation, called granules, are exceedingly small, and low contrast features require the best seeing conditions to consistently view. It is estimated that approximately 2–3 million individual granules cover the surface of the Sun at any given moment.

A granule is the top of a rising column of gas, originating deep within the convection zone of the Sun. In this zone, convection is the method of heat transfer. Hot plasma rises to the solar surface, releasing energy. Then, as the plasma cools, it flows back along what is called an intergranular wall or lane to the solar interior. The intergranular wall is what defines the shape of a granule; because the temperature of the material flowing back is cool, the wall appears dark. Each granule typically has a diameter of 1–5 arc seconds, with the average being about 2.5 arc seconds. These are tiny features on the Sun, but if placed on Earth the nearly 1100-kilometer diameter would consume the landmass of some nations.

There is no uniformity to the brightness of granules; some appear dull and others appear relatively bright. The lifetime of a single granule is 5–10 min. The larger granule is found to have a longer lifetime, with physical changes becoming apparent after only a minute or two.

More visible when near the center of the solar disc, the granulation is harder to see as the solar limb is approached. A telescope having a minimum aperture of 125 mm will be necessary to effectively study granulation. I recall my first observation of solar granulation, made with a 150 mm Newtonian telescope. The sky was unusually calm and the pattern stood out particularly well that morning. Even at a low magnification of 25×, the grainy texture seemed to engulf the entire disc of the Sun. Because green light enhances the contrast of solar granulation a supplementary filter is helpful for observation.

The short life span, combined with the necessity of near ideal seeing conditions, make intense study of solar granulation a difficult proposition for the amateur astronomer. Sensitivity to alterations in shape, size, and brightness within a several-minute time span is required. These criteria can be met by using video photographic techniques, which capture hundreds to thousands of images over the estimated lifespan of a granule. By culling the finest images from a video record, the observer can assemble a series of photographs or even a time-lapse movie depicting the brief appearance of a granule (Figure 4.3).

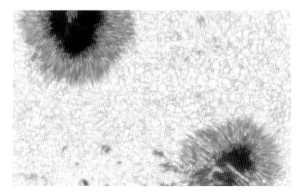

Figure 4.3. Photospheric granulation. Active regions 9600 and 9601 are imaged here at the *upper left* and *lower right*. Between and surrounding the two sunspots may be seen the kernel-like appearance of individual granules. Superb seeing conditions are required to spot the granulation. This image was captured September 3, 2001, by Art Whipple using a 203 mm f/10 Newtonian.

Faculae

Often located near and surrounding a sunspot will be a luminous, cloud-like patch or venous streak of material called a facula. Facula (plural is *faculae*) means literally a bright point. Although all sunspot groups are linked to faculae, not all faculae have attending sunspots. Facular regions are the precursor to sunspot groups. Should a magnetic field be too weak to allow a sunspot to form, only the facula will remain. A sunspot is a large, dark-looking region of magnetic strength; the facula is a magnetically weak, thin, granular-sized magnetic region. One characteristic of stifled convection, including those of lesser strength, is an apparent depression of the solar surface; the Wilson effect is an example of this aspect in a sunspot. A facula results from the light emerging through the "sidewalls" of this so-called depression.

Because of this scattering of light, a facula is seen as a bit brighter than the surrounding photosphere. Near the solar limbs, a facula will stand out when contrasted against the limb darkening, while in the solar continuum faculae are increasingly difficult to see as you approach the center of the Sun. An objective filter favoring blue light or a supplementary filter transmitting green/blue will accent facular contrast, permitting a feature to be seen closer to the center of the Sun.

The lifetime of a region can last for several solar rotations. An early warning system of sorts exists for an observer monitoring the emergence and growth of new facular regions. By maintaining a watchful eye toward these regions, a new sunspot group may be seen in its earliest stage of development.

Because of the relationship to sunspots, faculae occur primarily in the sunspot zones, about 35° north and south of the equator. A facula does appear occasionally outside these zones and near the polar regions of the Sun. Polar faculae differ from ordinary faculae in sizes and lifetimes. Small (granular-sized) point-like or elongated areas, polar faculae last from minutes to a day or two at most. The brighter examples will have a longer lifetime. During the minimum of a solar cycle, polar faculae are more frequently observed than during maximum.[1] The small area of a polar facula requires a minimum aperture of 100–125 mm used during fine seeing conditions. Solar projection screens are seldom the means of observing polar faculae; any ambient light falling on the screen extinguishes their low contrast appearance, and consequently direct observation with enhancing filtration is preferred (Figure 4.4).

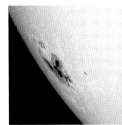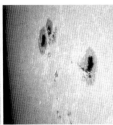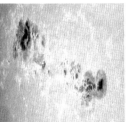

Figure 4.4. Faculae can appear orphaned or attached to a sunspot group, as shown in these images acquired with a 125 mm aperture refractor, mylar objective filter, and a Wratten #58 green filter. From *left* to *right* are AR0808 (2005), AR0464 (2003), AR0095 (2002), and AR9596 (2001). Jamey Jenkins.

Pores

On occasion, the intergranular wall may appear darker than usual, or a granule may look darker than its neighbors, or even be missing; there exists a natural variation of brightness within granules. Differentiating between these granular happenings and a feature called a pore can be an art. There is also an art to separating a pore from a sunspot because any sunspot lacking a penumbra may be a pore.

The pore is a tiny structure, with a diameter from 1 to 5 arc seconds. The average example will measure somewhere between 2 and 3 arc seconds. A pore will be darker than the granules just mentioned but brighter than the umbra of a well-developed sunspot. The lifetime of an individual pore varies widely. The smallest have been known to form and dissolve in a few minutes while other, larger specimens, may be visible for a number of hours. Size has a direct relationship to longevity. And the longer a large pore survives, the greater the likelihood that it will develop into a sunspot. Pores are often located near an existing sunspot group, or are found in some isolated location within a facular region. The formation of a pore is the result of the Sun's magnetic field extending upward from the solar interior and restricting convection on the photosphere. Rarely obtained but interesting to see, as with granulation, would be a series of images or a time-lapse movie produced by the amateur astronomer depicting the birth, development, and decay of pores (Figure 4.5).

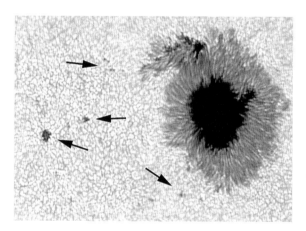

Figure 4.5. Pores are arrowed to the *left*; dark-appearing granular material is arrowed to the *right* in this image obtained by Eric Roel on August 12, 2006.

Sunspots

The most noticeable white light feature exhibited on the Sun is the sunspot. Looking like a blemish on the smooth, near-perfect photosphere, a sunspot will stand out in stark contrast to the surrounding surface. At low magnification, a sunspot gives the appearance of a light gray material, the penumbra, surrounding a much darker mass, the umbra. Several of these doughnut-shaped smudges may be clustered together in a section and attended by unattached portions of penumbrae and nearby pockmarks of pores. There may be a streak or river of whitish material, called a light bridge, seen crossing the darker central portion of the sunspot.

The shapes, sizes, and lifetimes of sunspots are as varied as the stars in the sky, although they consistently develop within about 35° on either side of the equator. This region is known as the sunspot zone. A symmetrically shaped spot demonstrates the cavity or depression effect experienced in a magnetically influenced active region by what is called the Wilson effect. As a round-shaped sunspot nears the solar limb, the exterior zone, called the penumbra, seems to increase in width on the limb side, more so than the side closest to the center of the Sun. The logical conclusion from this observation is that the sunspot has a concave or saucer-shaped depression to it. Studies indicate that this depression is illusionary, the result of gas becoming more tenuous or transparent in the magnetic field found in a sunspot, thus permitting the observer an opportunity to see deeper into the photosphere and giving the impression of a depressed profile.

Sunspot evolution takes a generally repeatable course that follows this basic scheme. Bright faculae will develop up to about 7–14 days before a sunspot will appear. Within the facular region, many pores will begin to develop. A number will be about the same size as the solar granulation and will decay and then disappear. Some pores will become larger and darker, becoming as dark as the typical sunspot umbra; this we call an umbra spot. This is where most sunspots end development, only to decay in a short time. If a sunspot is to continue to grow, a coarse penumbra will begin to appear. The penumbra may become a complex affair, with islands consisting of darker umbrae and bright points within its field. At this time we have a well-developed sunspot, likely to be accompanied by similar bodies, each in an evolutionary cycle that is associated with its neighbors.

Sunspots are often the focus of the amateur solar observer's white light investigations. The morphology (study of the changing appearance) of the Sun and the gathering of statistical data consume much of an observer's time. Indeed, it offers a unique opportunity for the curious to understand how magnetism is important, to both the Sun and life on Earth.

Sunspot Umbrae

The darker central component of a developed sunspot, called the umbra, if placed alone in the sky, would be brighter than any of the stars in the night sky. Umbrae appear dark only because of contrast with the overwhelming brightness of the surrounding photosphere. There exists a close relationship between the darkness of an umbral region and its magnetic strength and temperature. This is easy to understand when we consider that sunspots are the result of less convection on the photosphere; a stronger magnetic field creates less opportunity for convection and therefore a cooler, darker-appearing region.

Close examination of a sunspot indicates that the umbra is not uniform in brightness or color. In fact, an umbra is formed of dark granules, tiny bright points, and material exhibiting an in between intensity called umbral dots, not to be confused with umbral spots. Umbra within a large spot contains interior areas of darker intensity, disconnected from one other by slightly lighter regions.

Two photographic techniques are readily available for the amateur observer to study these phenomena. By creating a diaphragm that masks the surrounding photosphere and allows only the light from a sunspot's umbra to reach the camera, exposures can be obtained that reach deep enough to show the detail inside an umbra. The granular nature of an umbra then becomes apparent, an observation made difficult visually because of the insignificant differences in brightness between umbral details coupled with the limiting factors of atmospheric seeing. Secondly, the technique of creating an isophote map from a high-resolution photo reveals regions of differing intensities inside a sunspot. You may have seen this technique applied to an image of a comet visible in the night sky. The purpose of creating the isophote, in that case, is to illustrate the gradient nature of the comet's nucleus, coma, and tail. A sunspot isophote helps to identify a number of interesting features related to the sunspot, including the inner and outer bright rings, a weak light bridge, and the core (the point of minimum brightness and coolest temperature) within an umbra.

Initially, a quick peek at a sunspot's umbra indicates a very dark gray or almost black area. Inspection with a telescope that has no filtered bias toward color and is neutral across the spectrum reveals that an umbra actually is composed of black to a subtle yet deep reddish-brown color. The Herschel wedge or Baader Astrosolar film provides a neutral color view. Serious interpretation of sunspot umbra, like most solar studies, requires a telescope aperture of 125 mm or greater. With this instrument, sufficient resolution is available to show umbral detail, particularly if the above photo techniques are engaged.

Sunspot Penumbrae

Where an umbra is the darker central component of a developed sunspot, the penumbra is the lighter, grayish outer region surrounding the umbra. A rudimentary penumbra often begins forming from the intergranular material surrounding a newly developed umbra. In a spot with a large umbra, the rudimentary penumbra will evolve and develop structures of dark penumbral filaments that radiate about the umbra like fine threads. These filaments are magnetic in nature, having a similarity to granules in their convective characteristics. Between the dark threads are brighter regions called, penumbral grains. Superb seeing conditions and a resolution greater than 1 arc second is required to distinguish filaments within a penumbra.

A mature sunspot will generally have a penumbra that is symmetrical. Somewhat less seen will be an irregular penumbra, which has been mutated by complex magnetic fields. This type of penumbra will inundate the sunspot group with filaments of varying widths throughout the group. Independent islands of penumbrae may also be separated from the umbra. This condition is infrequent and rarely extends past a day in length; an observation obtained of any change in appearance would be noteworthy.

Within well-developed penumbra are found islands of dark umbral material that are only slightly larger than pores. Also in the penumbra can be found regions of material as bright as or brighter than the photosphere. These dark and bright regions are known to undergo rapid changes and should be observed closely. Occasionally, a bright region may become elongated and fade, transforming into filaments or growing larger still and becoming a light bridge.

A light bridge is fundamentally defined as any material brighter than an umbra that also divides an umbra, often times dividing even a penumbra. An older, mature sunspot may contain a thick light bridge, often appearing much like the photosphere has spilled into the sunspot. It is the younger sunspot that contains the thin, streaky, intense light bridge. Under close scrutiny, the granular appearance of a light bridge can be seen. The lifetime of the thin variety can be less than 24 h; the larger form, however, may last more than a week. When a mature sunspot develops a large, massive light bridge, the spot is in the downward slide of its life cycle.

Between the umbra and penumbra is a rough region where penumbral filaments have the appearance of extensions of the umbra. At this point, sometimes seen is a brightening within the penumbra called the inner bright ring. This occurs because filaments are brightest near the umbra, increasing in darkness the nearer they reach the exterior edge of the penumbra. Another feature visible is a brightening and aligning of the granules encircling the outer edge of a sunspot, beyond the penumbra. This is the outer bright ring (Figure 4.6).[2]

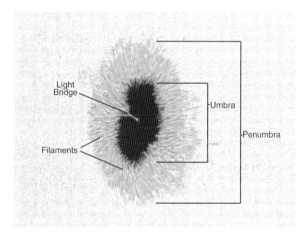

Figure 4.6. Parts of a sunspot. In this image of a symmetrically shaped sunspot, the filaments are seen aligned in a radial fashion indicative of the magnetic nature of the region. A small light bridge is developing, which will eventually divide the umbra into two parts. Eric Roel.

Sunspot Groups

A sunspot is not always a solitary creation. It tends to develop a nearby network that contains faculae, pores, and other spots. These features are all interconnected by a magnetism originating in the solar interior. A clump of sunspots is called a sunspot group.

Groups develop through an observed pattern that follows this route. Several pores may form a small clump, confined to a space less than 10° of solar heliographic area. These pores will darken within 24 h, becoming umbral spots separated into two distinct concentrations. In a few more hours, in each concentration, development will continue until a small sunspot is formed. The concentrations are termed leading and following, the leading spot being the more westerly of the two.

Evolution often terminates at this point with the dissolving of the concentrations within a few days. However, if a group is particularly stable and continues to evolve, a penumbra will develop about the leader spot, with penumbral material growing shortly thereafter around the remaining spots. The two concentrations of sunspots will now begin to separate from each other by at least 3° in solar longitude while they realign themselves, rotating relative to the solar equator in an east-west direction. The leading sunspot will have a magnetic polarity opposite that of the following sunspot; this is called a bipolar sunspot group. Should the clumping consist of only a single concentration, with its members confined within a 3-degree area, the result is known as a unipolar sunspot group.

About the middle of the second week of development a group usually reaches its maximum in area and number, to be followed up to a month later with the signs of decay. Pores and small spots will begin to dissolve. The following spot will divide and fade until it has disappeared. During these happenings, the leading spot becomes symmetrical in shape, possessing a uniform penumbra. Gradually the spot will have shrunk away, leaving only the faculae, also soon to disapate.[2]

At the height of development, a complex sunspot group can consume hundreds of millionths of a solar hemisphere. One millionth of a hemisphere is equal to 5 s^2 of arc as seen on the face of the Sun, when corrected for foreshortening as measured away from the center of the disc.[3] Extending over 15° in length and containing in excess of a hundred individual spots, a large group is occasionally the progenitor of flares, sometimes so powerful that they become visible in the solar continuum. This event, referred to as a white light flare, is rare and is worthy of special attention by the observer.

Classification of Sunspot Groups

Scientists have an innate desire to study, organize, and explain their subjects. Over the years astronomers, in order to classify the developmental stages of a sunspot group, have created various schemes. These classification systems are used to describe a group based on its measured properties, such as magnetism for the Mount Wilson classification system or for others on the visual appearance of the group.

The most useful visual classification scheme devised until the later years of the twentieth century was the Zurich sunspot classification system, by Max Waldmeier of Zurich Observatory. The Zurich system relied on nine classes named A, B, C, D, E, F, G, H, and J to label the various stages of sunspot development. This was a beneficial plan for identifying where a sunspot group fell in its life cycle, but for the practical purposes of the twentieth century, it fell short. Spacecraft and modern communications equipment required an early warning of any particularly destructive solar flares. A more reliable system with better flare prediction than the Zurich classification provided was necessary.

A flare is normally best seen in the narrow slice of monochromatic light in which the feature is in emission, that is, where it glows the brightest. Narrow band filters that isolate light from the H-alpha atomic line of the solar spectrum are quite effective when observing a flare. Sometimes, an extremely energetic event becomes so intense that it spills into the solar continuum, becoming visible in white light. Such events can be particularly chaotic to our local space weather.

Predicting when and where solar flare activity was likely to happen took a step forward in the 1960–1970s when Patrick McIntosh created an extended version of the Zurich sunspot classification system.

McIntosh revised the original nine-class Zurich system to a seven division one, similar to the older Zurich, but with the G and J classes omitted. Two sub-classes were added describing the penumbra of a group's largest spot and the distribution of spots within a sunspot group. The additional information derived from the three-letter McIntosh classification system made it adequate for flare prediction, giving professional and amateur astronomers an indication of when and where possible flare activity might happen.

To classify a sunspot group using the McIntosh system, an observer must inspect the group visually and determine where it falls within the descriptive language given for each code letter. For example, upon inspection an observer finds that a sunspot group is "unipolar with a penumbra" (H), that it has a "small, round penumbra having a diameter of 2.5 heliographic degrees or less" (s), and that it is "an isolated spot" (x). String these three letters together and the McIntosh classification is "Hsx."

Practice makes perfect is an adage that applies when classifying a sunspot group. New observers should make comparisons between the classification they determine and the classification posted on the Internet by professional observatories. Mees Solar Observatory in Hawaii provides sunspot classification facts daily. This activity will hone your observing skills, while developing an understanding of a sunspot's lifecycle. For the solar observer intent on seeing a white light flare, an understanding of the McIntosh system is vital to knowing where and when to look (Table 4.1, Figure 4.7).

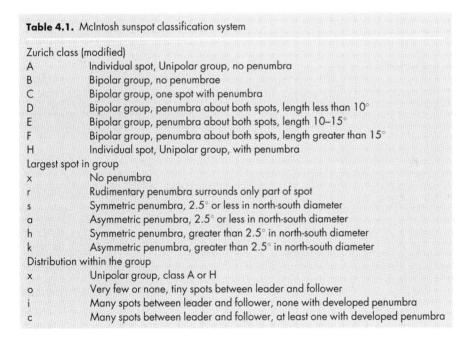

Table 4.1. McIntosh sunspot classification system

Zurich class (modified)

A	Individual spot, Unipolar group, no penumbra
B	Bipolar group, no penumbrae
C	Bipolar group, one spot with penumbra
D	Bipolar group, penumbra about both spots, length less than 10°
E	Bipolar group, penumbra about both spots, length 10–15°
F	Bipolar group, penumbra about both spots, length greater than 15°
H	Individual spot, Unipolar group, with penumbra

Largest spot in group

x	No penumbra
r	Rudimentary penumbra surrounds only part of spot
s	Symmetric penumbra, 2.5° or less in north-south diameter
a	Asymmetric penumbra, 2.5° or less in north-south diameter
h	Symmetric penumbra, greater than 2.5° in north-south diameter
k	Asymmetric penumbra, greater than 2.5° in north-south diameter

Distribution within the group

x	Unipolar group, class A or H
o	Very few or none, tiny spots between leader and follower
i	Many spots between leader and follower, none with developed penumbra
c	Many spots between leader and follower, at least one with developed penumbra

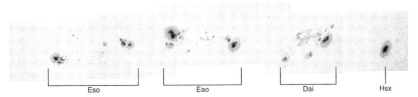

Eso   Eao   Dai   Hsx

Figure 4.7. Stretched across the Sun on July 9, 2000, were four sunspot groups of differing McIntosh classifications. Art Whipple.

White Light Flares

One of the most energetic but least glimpsed phenomena in the solar system is the white light flare (WLF). Becoming visible as a temporary brightening near a well-developed sunspot, the WLF takes on the appearance of patches, points, or bands of light about 50% brighter than the surrounding photosphere. The first solar flare seen was a WLF on the 1st of September 1859 by Richard Carrington, and confirmed by Richard Hodgson.

Solar flares are the manifestation of intense activity occurring in the chromo-sphere, the result of stress released between conflicting magnetic fields. A flare is frequently seen in the light of H-alpha beginning as a bright point or two, growing during a time period of minutes to several hours in brightness and size, and sometimes developing into a "ribbon-like" manifestation. If a flare becomes sufficiently active during its peak release of energy, it can become visible in the integrated light of the solar continuum. This peak is typically noticeable for less than ten minutes. Being in the right place at the right time can be crucial for an observation. This author has been observing the Sun on a more or less regular basis since 1990 and have had one occasion during this time of catching a truly brilliant WLF.

My WLF observation happened unintentionally, on November 24, 2000. Ordi-narily, little solar observing is done at this time year; during November the Sun is near the horizon. Consequently my observing window was less favorable. Regard-less, this day I was scanning the solar limb with a refractor fitted with a narrow band filter centered on the H-alpha line hoping to see any prominence activity that morning. After sweeping the west limb, I shifted the telescope across the face of the Sun to scan the east limb. As the telescope tracked, my attention was caught by a pair of bright patches within the penumbra of AR9236, an Eko (McIntosh class) sunspot group near the center of the solar disc. A quick peek through a telescope with a safe white light filter confirmed the patches had indeed become a WLF. In this case it was a matter of being in the right place at the right time. The flare lasted about another 2 min, having followed the usual pattern of rapid brightening to a peak and then slowly fading from view. Unfortunately, this brief time (1508–1510 UT) didn't permit me to photograph the event. Later, I discovered that fellow amateur astronomer Art Whipple had been observing that morning and recorded the flare with his equipment.

Although a truly bright example is rare, astronomers are convinced that a less intense variety of WLF can be seen more frequently. The key to spotting a WLF is being educated on when and where to look, being methodical in your approach, and optimizing the telescope for the search. Sunspot groups that have developed into McIntosh classifications of D, E, and F are noted for producing solar flares. The opportunity increases for those groups that are of the sub-classes ki and kc, though certainly not limited to these. Groups of Dai, Dso, and Hsx have, on occasion, sported flares. Begin by knowing the class of groups currently visible on the Sun. Pay close attention to irregular or detached penumbrae within a sunspot and areas of clustered spots between the leader and follower. As illustrated by my story, a white light flare can be a very transient phenomenon that requires excellent timing for its discovery. An occasional look at the Sun will hardly lead to success in spotting a white light flare. Although most of us cannot monitor a sunspot group on a continuous basis, it is possible to take a few moments throughout a weekend morning or afternoon, as when working in your backyard or garden, to scan a suspect group for WLF activity. A bit like visual comet hunting, this will eventually lead to a find, in this case the discovery of a WLF.

Optimizing the telescope for a white light flare search is as simple as being selective about the solar filtration. Because only the brightest of white light flares are visible on a solar projection screen, use direct viewing instead. The idea is to increase the contrast between the WLF and the surrounding photosphere. For observation, a safe visual density mylar objective filter with a blue transmittance is necessary. The mylar filter biased toward blue light will allow the continuum light from a flare in emission to pass, increasing its visibility. You can take this a step further by mating a weak blue eyepiece filter with a visually safe mylar objective filter to further increase contrast of the WLF.

If you search for WLFs by using video or photographic methods, try using a narrow band filter, passing 10 nm or less of light centered around 430 nm in the solar spectrum. This is the G-Band and is the location of spectral lines that go also into emission during a flare. This technique will dramatically increase the chance of catching a WLF, but the transmittance characteristics of this narrow band filter necessitate using thin photographic density filtration on the telescope. The increased amount of blue light with this technique is not safe for visual observing; therefore, only video or photographic observations are permitted (see the earlier chapter on safety).

When you see a WLF, note to the nearest second if possible the time of both the beginning of the observation and its disappearance, where on the Sun it appeared, and its relative brightness compared to the photosphere. Photograph the event or make sketches showing any changes in appearance. Lastly, report these observations to the appropriate organizations (i.e., A.L.P.O. Solar Section, B.A.A. Solar Division).

Heliographic Coordinates

To the naked eye or through a telescope, excluding limb darkening, the Sun is a flat disc, with features parading across it from east to west. The reality, however, is that the Sun is a globe-shaped spherical body suspended in our daytime sky. Foreshortening, the compression of regions beyond the center of the Sun, squashes noticeably the appearance of features nearing the limb.

Geographers on Earth created a system of latitude and longitude to define locations on our planet. Users of a GPS device are able, through communication with satellites, to determine a precise position within this imaginary grid of latitude and longitude. For Earth-bound residents knowing exactly where they are, the distance separating them from where they are going, and the precise location of where they are going is vital. Commerce and convenience run on a schedule in the modern world.

The solar observer, for analytical purposes, also requires a system of latitude and longitude that can be conveniently translated to the solar disc. Defining the position of a new active region for identification purposes may be necessary, particularly if the face of the Sun is crowded with an abundance of sunspot groups. Any study of a sunspot's motion relative to the photosphere or other spots requires such a system of reference points. This can be a complicated affair given that the Sun rotates differentially, has no well-defined permanent point of reference, and that Earth is continually in orbit around the Sun, creating differing views of a tipping and nodding body. Challenging? Yes, but there is a convenient method for today's solar observer of determining positions on the disc of the Sun.

Defining a position on the Sun requires knowing three parameters from the date and time of the observation that define the tipping and nodding of the Sun as seen from Earth. The first parameter is the extent of "nod" that the solar poles have taken during our yearly orbit. This is called Bo and represents the changing heliographic latitude at the center of the solar disc. The extreme nod of the northern hemisphere toward Earth, $+7.3°$, occurs early in the month of September. Conversely, during the first week in March, the southern polar region of the Sun is visible, as the latitude at the center of the disc becomes $-7.3°$. There are only short time spans each year when Bo is $0°$ and we are looking squarely at the center of the Sun's disc; this is during brief periods in June and December.

The second parameter needed is termed P for position angle. The P represents the amount of displacement the north rotational axis of the Sun has relative to the rotational axis of Earth. The amount of this offset or tip varies throughout the year by a total of $52.6°$ ($26.3°$ on either side of celestial north). When the solar north pole is tipped toward the east of celestial north, values are given a positive (+) sign. The maximum eastern tilt occurs in the early days of April, with a reading of $+26.3°$. As we progress in our yearly orbit, the Sun begins tipping back toward the west until its north-south axis is aligned with celestial north-south in early January, at this time, $P=0°$. Beyond this date the values of P begin to assume a negative (−) sign, reaching a maximum westerly tip of $-26.3°$ near the beginning of April. The tipping motion reverses again, with the north solar pole edging toward the east, passing through celestial north again, and a P of $0°$ near the first week of July.

These two values, Bo and P, from the date of an observation are all that is necessary to accurately fix the latitude of a feature on the Sun. To find the longitude of a feature requires defining the third parameter, an imaginary line drawn from the north solar pole to the south solar pole, the Sun's axis of rotation. This line is called the central meridian (CM) of the Sun. Solar astronomers measure longitude using the central meridian from the day of observation as the reference point. The longitude of the central meridian (abbreviated Lo) is $0°$ at the beginning of each new solar rotation, as determined by the Carrington rotation system. Heliographic longitude is measured increasingly from east to west, and because the Sun rotates from east to west, longitude at the CM decreases with time, progressing from $0°$ to $350°$ to $340°$, and so on. Lo decreases by $13.2°$ per day, which reduces to an hourly $0.55°$ decrease.[1] By interpolating the Lo of the central meridian using the above figures and information derived from a known daily table, plus measuring a feature's distance from the meridian, longitude of the feature can be acquired.

Certain tables, called Ephemerises, are published yearly that list the daily (usually at 0-h UT) orientation of the Sun for the factors P, Bo, and Lo. Several sources of this information include *The Astronomical Almanac* and the *Observer's Handbook* by the Royal Astronomical Society of Canada. Internet references include a yearly Ephemeris, published by the Association of Lunar and Planetary Observers (A.L.P.O.) Solar Section on their web page, or just type "Sun Ephemeris" into your search engine for a listing of numerous other web resources. Brad Timerson of the A.L.P.O. group has provided a daily solar Ephemerides through the beginning of 2012 that appears at the end of this book.

Recording Positions on a Photograph

An observation for measurement or reduction generally falls into a hard-copy format of either a whole disc (WD) drawing or photograph. Sketching at the telescope, while a relaxing activity, has a limited value for the modern solar observer. Although suitable for determining a group's position via solar projection (see Chapter 5), drawings tend to be only as good as the artist when an accurate rendition of a feature is required, particularly if a method of observing other than projection is used. Given the dynamic nature of the Sun and the rapid changes some features go through, a better choice for an accurate and permanent record is photography. The availability of a home computer system and the digital camera makes pencil and paper a hard sell when it comes to recording solar observations, other than for an occasional notation.

A whole disc photograph of the Sun is easily measured, providing a heliographic position accurate to about $1°$ near the center of the disc and, because of foreshortening, slightly less approaching the limb. Measuring a WD photo of the Sun can be accomplished mathematically or through the use of an overlay called, a Stonyhurst Disc. The overlay method of measurement is probably less tedious than the otherwise complicated mathematical calculations; therefore, it is a preferred technique.

A Stonyhurst Disc is a template showing the lines of latitude and longitude for a specific value of Bo. A typical set consists of discs in single degree increments from $0°$ through $7°$. By "flipping" the grids head for foot, the negative values of $-1°$ to $-7°$ become available. Stonyhurst Discs are obtainable through some observing organizations, but the easiest way is to download them through an Internet source and print the discs from your computer. Several options are available, including the resource provided by British amateur astronomer Peter Meadows (www. petermeadows.com). This site provides links to discs in several formats and of a variety of diameters to meet any amateur's needs. Meadows has also assembled excellent tutorial material and has available a freeware program, called Helio, which can be used to determine sunspot positions from X to Y coordinates on the solar disc, as well as providing the daily parameters for P, Bo, and Lo (Figure 4.8).

The WD photo of the Sun to be measured for heliographic coordinates must be of the same diameter as the Stonyhurst Disc. The larger the diameter, of course, the greater the accuracy that can be achieved. Standard disc size for a WD photo is 150–180 mm. Print the Stonyhurst Disc downloaded from the Internet on a clear transparency media or a translucent paper. Because the overlay is to be put above a photo, it could be difficult to see solar features if the disc is printed on an opaque sheet of paper.

The tools necessary to measure a photo include: a Stonyhurst Disc having the correct value of Bo for the day of observation, a protractor, a ruler, a fine tip ink pen, several paper clips or scotch tape, a pushpin or needle, and the values of P and Lo for the day and time of the observation. After you have measured a few photos, you may very well determine a shortcut or two that will save you time and steps.

To begin the task properly, a WD photo must have indicator marks that align within about $1°$ of celestial north-south or east-west. This can be obtained when the photo is secured by incorporating two pointer marks at some intermediate point of focus within the camera system, touching the opposing solar limbs in a N–S or E–W position. Some observers adjust the camera so that the upper or lower

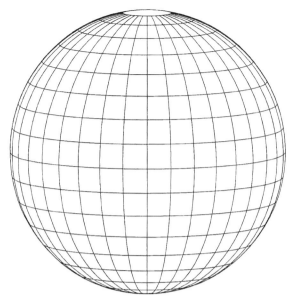

Figure 4.8. Stonyhurst Disc with a *Bo* value of 7°. Flip the template 180°, and the *Bo* = −7°.

edge of the frame is parallel to E–W; do this by watching a sunspot drift along the edge and rotate the camera appropriately. One good technique is to take an "alignment photograph" before snapping an actual observation photo. An alignment photo using the afocal method of photography is obtained by inserting an eyepiece having either a crosshair reticule or, better yet, a thin single wire at its focal plane dividing the field of view. Disengage the drive on the telescope and rotate the camera package to align the thin wire or crosshair to a drifting sunspot. When the wire is precisely positioned east-west, center the Sun in the frame and snap a picture or two showing the wire crossing the face of the photosphere. A print of this image at the same scale as the final observation photo serves as a reference for locating the true north and south poles on the Sun. If an alignment photograph is of suitable quality, it can be used for measurement directly. All hard-copy prints should be oriented north-up and east to the left, just as the Sun appears to the unaided eye.

Take the alignment photograph and find the celestial east-west points by dividing the hemisphere of the Sun at its center with a line drawn parallel to the image of the wire. Now draw a line perpendicular to that line passing through the center of the Sun. The Sun should now be divided into four equal quadrants, the vertical line defining celestial N–S and the horizontal line the celestial E–W points. Take the protractor and measure from the north celestial point on the Sun the amount of *P* given in the Ephemeris for the day of observation. Positive values are measured toward the east and negative values toward the west of celestial north. Draw tic marks on the ends of an imaginary line passing from the point of *P* through the center of the Sun and exiting the opposite hemisphere. The upper tic mark represents the position of the north pole of the Sun and the lower mark the south pole. Now superimpose and square up the alignment and observation photos, and, using the pushpin or needle, poke tiny holes through both photos at the N–S poles. Use the ink pen and ruler to redraw the tic marks representing the north and south poles on the observation photo.

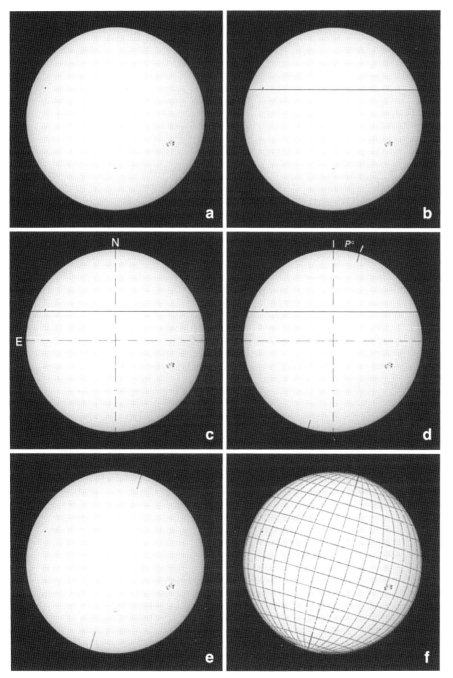

Figure 4.9. How to use a Stonyhurst Disc. Figure **a** is the observational photo from May 29, 2003, at 1530 Universal Time. Photographed just prior to this in **b** is the alignment photo with the wire positioned so that the spot located on its edge in the northeast drifts accurately along the wire. In figure **c** the alignment photo has been quartered by a *dotted line*, E–W is *parallel* with the wire, N–S is *perpendicular*, and celestial north and east are identified. In figure **d** the solar orientation for *P* is drawn with the use of a protractor; that location is transferred to the observation photo **e**. The Stonyhurst Disc with the correct value for *Bo* is aligned over the observational photo **f**, providing

The Stonyhurst Disc having the proper value of Bo is now positioned over the observation photo. Remember that a positive Bo causes the north polar region of the Sun to nod toward the observer; this effect will be visible in the latitude lines on the Stonyhurst Disc. When selecting a Disc from the set, pick the one with the nearest Bo value for the day of observation. For instance, if the Bo from the Ephemeris is +4.6, select the +5 Bo Stonyhurst Disc. Center the disc on the edge of the solar limb and rotate it until the central meridian longitude line (N–S) intercepts the north and south tic marks drawn on the photograph. Using paper clips or pieces of scotch tape, affix the Stonyhurst Disc temporarily to the photograph (Figure 4.9).

The resulting sandwich of photo and overlay depicts the location of the lines of latitude and longitude on the photosphere. Finding the latitude of a feature is rather straightforward; the equator is designated as 0°, and each line above or below represents a 10° increment. It will be likely that some measurements will have to be interpolated between lines; the ruler is helpful for this purpose. Latitudes north of the equator are designated as being either N or (+), and those to the south are S or (−). Solar longitude is measured from the central meridian and is presented in either of two forms, the Carrington system or as "relative." With the Carrington system, it is necessary to determine the longitude (Lo) of the CM at the time of observation. Computing the longitude of a feature is then accomplished by measuring its distance from the CM in degrees again each division on the Stonyhurst Disc representing 10°. It may be necessary at times to interpolate between the lines. Once the distance from the CM to the feature is known, add or subtract that distance from the CM longitude. A longitude to the west of the CM is a greater number than the Lo; therefore, add the feature's distance from the CM for its longitude. An eastern hemisphere longitude is smaller than the CM, so a feature's distance is subtracted from the Lo. The Carrington system is used for precise investigation of active regions.

The relative system simply utilizes a feature's distance in degrees from the CM, whose reading is accepted to be 0° at the time of the observation. Longitude east of the central meridian is designated as E, and longitude to the west as W. Relative longitude is convenient and is widely accepted, so long as the date and time of the observation are also noted with the heliographic coordinates.

References

1. *How to Observe the Sun Safely*, L. Macdonald, Springer-Verlag, 2003
2. *Handbook for the White Light Observation of Solar Phenomena*, Richard Hill, ALPO, 1983
3. *Solar Observing Techniques*, C. Kitchin, Springer-Verlag, 2002

Figure 4.9. (Continued) direct reading of a feature's solar latitude and longitude. The large group south of the equator and to the west has relative coordinates read as S07W42. In the Carrington system the coordinates are read as latitude 7°S, or −7° and longitude 200.7°. The longitude was determined by adding to the CM longitude at the time of observation (158.7°) the relative longitude of 42°; the result is the Carrington longitude.

Chapter 5

Recording White Light Observations

Observing Programs

An organized methodology is one route when studying a particular aspect of the Sun. But participation in an observing program is an excellent way to hone skills as a solar observer, become knowledgeable about the Sun, and even make a contribution to science. For example, my learning to recognize the quality of seeing conditions developed only after I began participating in the regular observing regimen of the sunspot-counting program administered by the AAVSO. In other words, having an organized plan to observing will enable you to learn more about solar phenomena and the morphology of the Sun, than you could have been taught elsewhere.

Several amateur organizations exist around the world for the purpose of coordinating the observations conducted by their membership. In the United States, the American Association of Variable Star Observers (AAVSO) has a division dedicated to solar observing. The primary task of the AAVSO Solar Division is to gather data from observers interested in sunspot counting. The Association of Lunar and Planetary Observer's (ALPO) Solar Section, another U.S. group, collects observations that are focused on the changing appearance of the Sun. Solar images from ALPOSS contributors are posted on the group's web page, and a communication network exists for observers through its Internet e-group. In the U.K., the British Astronomical Association also has an active solar section, coordinating work by amateur astronomers on that side of the ocean. A listing of solar organizations dedicated to assisting an amateur in his or her studies is located near the back of this book. Contact one or two, and inquire into their observing programs. It is an excellent way to learn, contribute to science via your hobby, and make new friends.

How does a person go about finding the right program? Well, usually, the program finds you. If you have a curiosity about the Sun, which you must certainly have to be reading this book, ask yourself these questions: "What is it that attracts me to viewing the Sun? Am I interested in the quantity of sunspots, their locations on the Sun, or do I tend to focus on the individual group and take notice of the changes occurring within it? Am I the type of individual that enjoys working with others on a project, or do I boldly go where few have gone before, a modern day Christopher Columbus, of sorts?" Answer these questions, and you will have an idea what direction to take. Sunspot studies tend to attract a greater number of participants. Unique features, such as polar faculae, lure the fewest observers. Whatever the motivation, if the Sun is your hobby, there are numerous avenues to

J.L. Jenkins, *The Sun and How to Observe It*, DOI 10.1007/978-0-387-09498-4_5,
© Springer Science+Business Media, LLC 2009

be explored, all of which prepare you to be a better observer and a contributor to the wealth of data being acquired about the Sun.

White light observing programs fall within two broad categories: statistical or morphological studies. An observing program that contributes statistical information could be the recording of the relative sunspot number (R), or documenting data on group distribution or sunspot classifications. A similar numerical study could be performed regarding facular activity, including data collection on the brightness levels of faculae.

The other category, morphology, is the recording of physical changes found in a solar feature. A morphology program could gather data on the evolution of a sunspot group. This might include a photographic record of the group's life cycle as it crosses the face of the Sun. Another project for the observer could be a visual patrol for the elusive white light flare. The possibilities are limited only by an observer's own creativity.

It's not possible to outline every area of study for the solar observer, so we will discuss programs that have proven successful in the past and are still being used. Consider your own interest, and then become involved in an observing program suited to your personal inclination. The satisfaction acquired through participation in a program will be rewarding, and the enjoyment you find in your hobby increased several fold.

Statistical Programs

The existence of sunspots has been known since early Chinese astronomers first glimpsed large naked-eye specimens of them many centuries ago. It was not until telescopic observations began about 400 years ago that the nature of sunspots and faculae became clearer, and the rotation of the Sun discovered. Galileo and Scheiner were major contributors to this field. Heinrich Schwabe, through his search for a planet orbiting between the Sun and Mercury, stumbled upon the periodicity of sunspots. Schwabe recorded sunspots only to be able to discount them, in the futile hope of discovering a new planet.

An actual statistical study of the sunspot phenomena began around 1848, with observations conducted by the Swiss astronomer Rudolph Wolf. Wolf was searching for a refined method of measuring solar activity. His first choice, rather than counting sunspots, would have been to ascertain the weight of activity by sunspot area, but to do so required equipment that he lacked. Therefore, he devised a scheme of counting the number of individual sunspots visible on the disc of the Sun and adding to this the product of the number of groups seen, as multiplied by ten. His logic behind the totaling of the two quantities, supported by decades of observational data was that independently, neither quantity alone was an accurate gauge of the true solar index, but combined they formed a true picture of solar activity.

It is through cooperation with other like-minded observers that the individual statistics average out, creating an accurate picture of daily solar activity. As discussed in the previous paragraphs, several organizations exist for the purpose of coordinating the data collection from observers. Some of these organizations offer unique programs, others maintain programs that collect identical data. Regardless, all are in the business of promoting solar astronomy and connecting amateur astronomers in a common cause.

Sunspot Counting

For the person seeking an activity that requires a minimal amount of equipment, sunspot counting is by far the most beneficial way to insert yourself into an organized program. A telescope fitted for projection or direct viewing of the Sun, a note pad, a pencil, and a timepiece are all that is required for scientifically useful data collection. The time spent at the telescope is minimal, requiring only several minutes per day to complete an observation.

The official sunspot number is an estimate, a compilation of observations made by a large number of independent observers. Several experienced amateurs have asserted that sunspot counting is more of an art than a science. This is true, but only because of the factors that contribute to differences in individual same-day counts, such as the observer's experience, seeing conditions, varying solar activity within a 24-hour counting period, and equipment. For instance, Wolf used an 80 mm refractor at a magnification of 64× for his sunspot counts. A larger telescope with higher magnification, under similar conditions, with an equally skilled observer could show smaller spots if they existed because of its greater resolution. And then, there is the issue of what is a small sunspot and what is a pore. The observer is not supposed to count a pore, but one person's interpretation of what is a pore could be different from another person's interpretation. As you can see, sunspot counting is at best an estimate obtained by the work of an artisan practicing a science.

The relative sunspot number, R, the standard established by Wolf over 150 years ago, is computed from the formula $R = 10\,g + s$. The total number of sunspot groups visible to an observer during a session is designated "g"; the total number of sunspots is "s." Computation is quite simple once the data has been obtained; it is often kept in a tabular form in the observer's notebook. A scaling factor, called "k," is often administered to an observer's submitted sunspot counts by the compiling organization. The purpose of k is to correct for the variables outlined above, that is, an observer's level of experience, local seeing conditions, and instrument differences. Another purpose of k is to link modern day observations with those from Wolf's time, so that continuity is maintained. The corrected relative sunspot number is computed as $R = k(10\,g + s)$.

Let me describe how an observing session unfolds for me; eventually you will probably develop a similar routine that works for you. First of all regular observing is important for polishing the skills required in this activity. Generating a sunspot count on only 2 or 3 days a month guarantees poor results over the long term. It is necessary to observe as often as possible, ideally each clear day. What works best for me is to have a scheduled time when I make my daily count. Usually mid-morning, before the lunch hour, is the time I'm able to set aside for solar observing. The daily sunspot count is always the first item on my agenda.

I begin by directing the telescope towards the Sun, either by watching the shadow created on the ground or by utilizing a pinhole solar finder. For direct viewing, a visual objective filter is preferred. If I'm using a projection device, the screen must be shaded from sunlight; otherwise, small, dim sunspots will be difficult or impossible to see. Since the majority of observers use direct viewing, I'll for the most part limit my discussion to that method. I record the sunspot count at the telescope on a pre-printed card, 100 mm × 150 mm (4 × 6 in) in size and containing a disc for sketching the approximate position of all groups on the Sun.

Space is provided for information regarding instruments, seeing conditions, and other pertinent notes. The sketch can be useful when locating a sunspot group at the next observing session that may have evolved in a way that makes it difficult to recognize. Solar rotation, as you recall, would have taken everything on the disc a bit farther to the west.

In Figure 5.1, an example of the pre-printed card is given from an observing session on May 29, 2003. With the refractor I use, a star diagonal creates the orientation of the image indicated in the drawing, with north up and west to the left. Your telescope may exhibit a different orientation. These are celestial directions and can be found by gently nudging the telescope N, S, E, or W and noting the displacement of the Sun. Select an eyepiece that yields a magnification suitable for viewing the whole solar disc. I often use a light red/orange filter on the eyepiece for sunspot counting. The purpose of the Wratten filter is to artificially darken umbrae and make tiny, faint sunspots stand out from the background photosphere. Carefully focus the telescope on a sunspot group. If no group is immediately apparent, use the limb of the Sun. Consistency is a particularly important factor with any statistical program. Since reliable results are only obtained after a long period of time, I never bounce back and forth between telescopes, and the only information logged is information I am absolutely certain of.

Scanning inside the sunspot zones, to about 35° on either side of the solar equator, I note all the groups I see, and sketch their rough locations on the card. A square or rectangle approximating an area of a group is used for this task. It may be necessary at times to "shake" the telescope slightly, to cause a small spot to become noticeable.

This is where the science becomes an art in solar observing. Familiarity with the McIntosh classification system is integral for recognizing what is an individual group and what may be two or more groups in close vicinity. Misinterpreting two

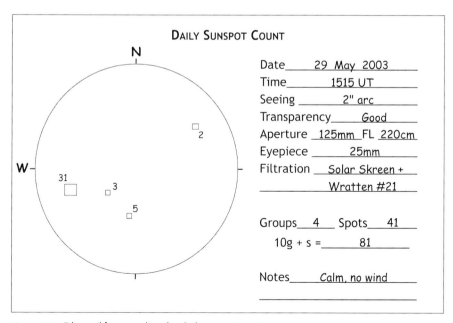

Figure 5.1. File card for recording the daily sunspot count.

groups as one will skew a daily sunspot count by 10, and if this occurs in several instances the count will be hopelessly inaccurate. Basically, I find it works in this manner: if a clump of spots doesn't fall within the guidelines established in the McIntosh system then I re-evaluate the clump, possibly defining it as several nearby groups. Having a card at the telescope with the McIntosh classes printed on it for reference is an excellent idea. If the pre-printed card system I use for recording observations appeals to you, listing the sunspot classes on the reverse side of the card would be advantageous.

Regular solar observing is an aid to determining group status. Why? Because the regular observer is a witness to a group's evolutionary process. There are days when I am unable to observe, and on those days I use the Internet to maintain a connection with activity on the Sun. But I refrain from "checking out" the Sun online if I plan on observing later that day. I don't want to influence my daily sunspot count by what I might see on the web.

Some observers use a "10-degree rule" when determining one group from another. This rule states that "any spot or group of spots that is at least $10°$ of heliographic latitude or longitude away from any other spot counts as one active region [group]."[1] Occasionally, two groups may form within $10°$ of each other, and the rule has to be bent to accommodate the situation. In most circumstances, however, the rule will apply, and groups will be separated by well over $10°$.

Having sketched the sunspot groups that are seen at a whole disc magnification, the eyepiece is then switched to one of $90-100\times$, and beginning at one limb I carefully scan the sunspot zones, searching for any other tiny, compact, or faint groups. Remember that a single isolated spot is considered a group; if one is swept up, include it on the sketch. Use a red/orange supplementary filter to help distinguish a spot from the background; a lighter shade than before may be necessary because a greater magnification results in a dimmer view.

Once I am confident that all the groups have been correctly noted, and still using a $90-100\times$ eyepiece, I start on one side of the Sun and inspect each group, counting individual sunspots. The total number of spots in each group is written beside its corresponding square or rectangle on the sketch, and then I move on to the next group. This continues until all the spots have been counted.

The inclination at this point is to ask, "What's a spot and what's not?" A granular feature can appear dull; a pore is darker than a granule, with an average diameter of $2-3$ arc seconds; and neither of these is counted as a sunspot. A sunspot umbra is darker than a pore and may be several arc seconds in diameter, up to a size large enough to be glimpsed with the naked eye. Any spot that is conjoined is counted as one; as soon as it separates and the pieces no longer touch each other, it is counted as two spots. A light bridge divides a sunspot. Until the spot is completely separated, consider it as one. Penumbrae and pieces of penumbra are not counted as sunspots; if it becomes difficult to distinguish small pieces of penumbra from umbra, use the brightness test by comparing the suspect spot with nearby known umbral features. If the brightness difference is significant, consider the suspect as a penumbra. If similar, count it as a spot. Years of observing have taught me that experience is the best instructor in the "art" of sunspot counting.

After all the spots have been counted – remember that during sunspot maximum the total can easily be over a hundred – I return to the starting point and repeat the count. It's easy to miss a spot here or there, so I correct the note card as needed. After the second confirming spot count, I fill in the remaining data on the card: the date, the Universal Time when the count was completed, an estimated interpretation of seeing and transparency conditions, and the data regarding instruments used. Note that the eyepiece identified is the one used to make the spot count, not the group evaluation. The relative sunspot number is calculated from my group and sunspot total and is entered on the card. The cards from all observations are filed until the monthly report form is due to be completed and submitted to the appropriate national organization.

An alternative to the above card method, one intended for a statistician wishing to have a more detailed observing program, is to maintain a log sheet on either a clipboard or in a notebook. A sample log sheet is provided in Figure 5.2,

Daily Sunspot Count   MONTH/YEAR _____ June ____ / ____ 2003 _____

	UT	S	T	Gn	Sn	Gs	Ss	G-total	S-total	R
01	1530	2"	G	2	12	2	12	4	24	64
02	1530	2"	G	2	15	1	6	3	21	51
03										
04										
05	1545	<1"	E	4	25	2	2	6	27	87
06	1540	2"	G	4	50	2	6	6	56	116
07										
08										
09										
10	1615	5"	F	3	87	3	59	6	146	206
11	1619	2"	F	4	83	3	69	7	152	222
12	1544	1"	G	3	51	3	89	6	140	200
13	1530	1"	G	3	30	3	85	6	115	175
14	1530	2"	G	3	9	3	42	6	51	111
15										
16										
17	1650	3"	G	2	14	4	29	6	43	103
18	1647	4"	G	2	15	4	26	6	41	101
19	1550	4"	F	2	17	4	57	6	74	134
20										
21	1626	2"	F	3	26	3	72	6	98	158
22	1645	2"	F	2	20	2	54	4	74	114
23	1640	3"	F	2	33	3	31	5	64	114
24	1635	2"	F	3	41	2	35	5	76	126
25	1700	1"	G	5	38	3	16	8	54	134
26	1725	1"	G	6	54	2	14	8	68	148
27										
28										
29	1538	3"	G	7	32	3	33	10	65	165
30										
31										
Monthly Total				62	652	52	737	114	1389	2529
Monthly Average				3.3	34.3	2.7	38.8	6	73.1	133

Aperture __125mm__ fl _2200mm_ Eyepiece fl__ 25mm __Magnification____ 90x ____
Filtration_____ Baader Visual Objective + Wratten #21 _____

Figure 5.2. Recording log sheet for the month of June 2003.

illustrating a typical month of sunspot counting. The monthly submission form provided by several national organizations is similar in format to this one, which I designed on a home computer. I print copies as I need them. You will notice that more information is permitted on the log sheet than on the card. Across the head of the page, columns are identified as follows:

UT is the Universal Time (at the completion of the observation)

S is the seeing conditions (defined by the arc second method)

T is the transparency of the sky: excellent, good, fair, or poor

Gn is the number of sunspot groups north of the solar equator

Sn is the number of sunspots north of the solar equator

Gs is the number of sunspot groups south of the solar equator

Ss is the number of sunspots south of the solar equator

G-total is the sum of Gn and Gs

S-total is the sum of Sn and Ss

R is the relative sunspot number, computed from $R = 10 \, g + s$

The monthly total and average are entered along the bottom for convenience. An average is based on the number of observation days, not the number of days in a month. The telescope data at the foot of the log sheet could be preprinted on each sheet since the same instrumentation is to be used daily. Note that the magnification sited on the form is the one used for counting individual sunspots, not groups.

It is possible to glean R from the cards or a log sheet and graphically illustrate it against time, as in Figure 4.1. Also available from either of these tools is a sunspot group statistic called the mean daily frequency (MDF). To calculate the MDF for the month in question, add the daily totals of sunspot groups and divide this by the number of days observed. In Figure 5.2 the MDF for the month is 6, calculated as the sum of column "G-total" divided by 19, the number of days with entries. Finding the MDF is about as simple as it gets with sunspot counting, it requires noting only the number of active regions containing sunspots visible on the Sun. When diagramed against time, the MDF will show the rise and fall of solar activity within a solar cycle.

A more complex study can be conducted by tabulating groups and spots based on the hemisphere in which they develop. The log sheet organizes this information in the columns Gn, Sn, Gs, and Ss. Collecting this data is important because the amount of activity occurring in one hemisphere can be substantially less than what is occurring in the other. The observer in another project could determine the latitude of groups in heliographic coordinates. When the latitude of sunspot groups are displayed against time, a unique pattern, called the butterfly diagram, results. A butterfly diagram visualizes the migration of sunspots from the higher latitudes towards the equatorial region as a solar cycle advances. The key to sorting out these detailed studies is recognizing whether a group is north or south of the solar equator. As you might recall, the Sun appears to nod and tip as Earth swings through its yearly orbit, changing the apparent position of the solar equator from month to month. Although celestial directions are readily discernible, the true solar orientation through an eyepiece could be a guesstimate, an unacceptable practice for this type of data collection.

One way to find the solar equator uses a Stonyhurst Disc of the proper Bo value, which is orientated over a whole disc photograph or drawing from the day of observation. When a whole disc photograph is measured, as outlined in Chapter 4, the equator is obvious. However, not every observer of the Sun is a photographer, and then the exercise must be done using a slightly different approach.

The Hossfield pyramid is ideal for making a position drawing that can be measured with a Stonyhurst Disc. To create the drawing I place on the projection screen a sheet of white paper, which has a disc bisected by two perpendicular lines. It is important that this "drawing blank" has the same diameter as the Stonyhurst Disc. A projected image of the Sun is adjusted to fit to the edge of the disc, and then the drawing blank is rotated until a sunspot accurately drifts along one of the bisecting lines. I then mark with a pencil the ends of the celestial east-west line. The other line perpendicular to the first represents celestial north and south. Mark the north and south points accordingly. An observer with an equatorially mounted telescope can leisurely follow these steps, but if the telescope is on an alt-azimuth or Dobsonian style mount, do work quickly or else field rotation, encountered as Earth turns, may skew the accuracy of the drawing.

Center the projected disc of the Sun onto the drawing blank and carefully mark the position of all the sunspot groups you see using either dots, short dashes, circles, squares, or rectangles. An artistic representation of a group is NOT what is needed for sunspot counting. The purpose of this activity is to determine where on the face of the Sun the sunspot groups are located. Once all the groups are identified, it is possible to obtain a count of the number of spots seen on the projection screen. It is preferable, however, to make a sunspot count using the direct viewing method. One reason for this is that tiny spots that are abundant in a well-developed group can become lost in the texture of the projection screen, altering the results. Additionally, contrast is improved when viewing the Sun directly through the telescope; spots become more apparent, particularly when supplementary filtration is used. My technique, therefore, involves removing the drawing from the projection screen at this point, switching to the objective filter, counting spots, and noting on the drawing the number of spots visible in each group.

Of course when an observation is completed, all pertinent information is jotted down on the drawing: date, time, seeing conditions, and so on. Later, using a protractor I find the position angle for the north pole of the Sun relative to the N–S line on the just completed sketch. Tic marks are penciled on the drawing for the correct value of P, and the Stonyhurst Disc with the correct Bo for the day of observation is positioned over the drawing. The equator becomes obvious, indicating which groups are located in the northern and southern hemisphere.

It is not essential to use a Stonyhurst Disc if a group is obviously positioned far away from the solar equator. However, groups do form within a few heliographic degrees of the equator, and then it becomes absolutely necessary to determine positions accurately. This happens frequently near sunspot minimum, because according to Spörer's law, sunspots of a dying solar cycle form in the lower latitudes.[1]

Polar Faculae

Another statistical program for a white light observer to examine is the daily monitoring and logging of polar faculae. This phenomenon is visible near the solar limb at heliographic latitudes of 55° or more. Appearing as small bright or elongated spots in the photosphere, the lifetime of a polar facula varies from a period of minutes to several days (see Chapter 4).

Direct viewing with at least a 100–125 mm aperture telescope during times of superior seeing conditions is perfect for this activity. The feature is usually tiny and of low contrast, making its observation difficult with a solar projection screen. An objective filter with supplementary filtration in the green region of the spectrum is recommended.

The goal of a polar faculae patrol is to document the number of faculae visible during an observing session. Since many of the faculae appear and disappear relatively rapidly, a great deal of time patrolling is not necessary. Begin by scanning at about 75–100× the solar limb above 50° latitude north and south, inward about a quarter of the solar radius. Sweep with an arc-shaped path from one side through the polar region to the opposite side of the Sun. Make a couple of passes to confirm any facula noticed, then switch to the other hemisphere and repeat the process. It is important to realize that during some months of the year, the north or south pole of the Sun will be tipped toward Earth. At these times one region favors observing over the other, and an imbalance in the hemispheric count is expected.

The log sheet for recording polar faculae can be an elementary affair, not much more than a column or two added to a sunspot record sheet. Provide a column to list the number of faculae visible in the north and a second column for faculae in southern regions – or provide just a single column for the total number visible. Add the columns at the end of the month and divide by the number of observational days to find the monthly number. Dedicated observers, through extended observing sessions, might try to document by visual means the lifetime of some of these events. That is, however, less of a statistical study and more of a morphological endeavor.

Morphology Programs

The study of the changing appearance of the Sun's features is called morphology. A detailed portrait of sunspots, a whole disc sketch or photo depicting sunspot positions, and a high-resolution photograph of a facula or granulation pattern are all examples of data acquired for the morphological study of the Sun. This is an area that many observers find very fulfilling and exciting. The old adage that a picture is worth a thousand words has never been truer than in the experience of the amateur solar astronomer doing morphology.

The purpose of any visual or photographic record is, in essence, to tell the story of an event. The astrophotographer that manages to secure even a single, sharp photo of the Sun holds in his hands a piece of time. This is where the value of our observation is found; because of the rapidly changing appearance of the Sun, all records are equally in demand. Each has a uniqueness all its own. A "photo series" is particularly valuable for illustrating the ever-changing aspects of the Sun. Some of the most telling series are the development of a sunspot group as it marches across the solar disc after first appearing from around the eastern limb. A cluster of small spots on the photosphere evolving into a giant sunspot group can occur quite rapidly and unexpectedly. For an advanced amateur, the possibilities also include the production of time-lapse movies, bringing to life what would otherwise be only short vignettes in the solar observing experience.

Drawings or Photographs?

As we've said before, this book does not strongly advocate the drawing of solar features. Admittedly, there exist times when pencil and paper are the sole means available for recording an event. And for notational purposes, a sketch serves its aim well. In the twenty-first century, however, the availability of many recording devices (film and digital cameras, computer web-cams, etc.) as well as the proliferation of the home computer has turned drawing into more of a therapeutic exercise than one producing scientifically useful results.

A quality photograph (or image, in the modern vernacular) is a far more accurate representation of a feature than a drawing, unless of course you have the skill to create a technically superior sketch. Remember that solar features are constantly evolving, and it takes time to produce a detailed drawing; often a feature will have changed its appearance, sometimes markedly, from the time a drawing was begun until it is finished. A photograph on the other hand, effectively freezes time at the instant it was taken; an exposure of the Sun is but a blink of the eye. Even the owner of a simple point and shoot digital camera can produce reliable results that rival the best photos obtained in amateur circles only a generation ago. Thus, an amateur interested in solar morphology is far better off to expend his or her energy securing photographs of the Sun, rather than attempting to draw those features observed. With that as a basis, in the following pages we will present several photo programs an amateur solar observer could pursue. To avoid repetition, detailed instruction of photographic techniques will be provided in a later chapter entitled, *Solar Photography*.

Whole Disc Photos

Around the world, professional solar observatories try to obtain a daily whole disc photo of the Sun in white light as well as the Sun at other selected wavelengths. These photos are used to analyze the motion and drift of active regions. Of course, not every observatory has ideal weather or seeing conditions year round, and this glitch provides the window whereby an amateur astronomer can contribute to the daily data collection – that is, by participating in an observing organization's photography programs. By filling the gap left in the professional ranks, the amateur solar photographer is making a significant contribution to solar astronomy and enjoying a hobby, too.

Obtaining a whole disc photograph of the Sun in white light is similar to shooting a picture of the Moon. In fact, it is often suggested to novice solar photographers that they start out by attempting to first image the phases of the Moon. After learning the necessary skills to obtain a sharp, explicit photo of Earth's natural satellite, graduating to the Sun will be an easier task. The Moon offers such a rich landscape, with contrasting views near the terminator, that focusing a camera on it is literally a snap. Exposure times for the Moon and Sun are similar, since most visual filtering systems effectively dim the Sun to about the brightness of a full Moon. It is important to have a degree of success under your belt before attempting a challenging target like our Sun, whose features tend to be of low contrast and sometimes difficult to see through turbulent daytime skies. Photographing the Moon can achieve that preliminary success.

Pursuing a whole disc photography program requires an observer to ideally photograph the Sun on every clear day. If you are a weekend observer don't expect to be able to contribute systematically to this program. However, once a routine has been established, it only takes several minutes a day to obtain the necessary photo of the Sun. Whole disc photography is well suited to an observer who can make the long-term commitment but has time for only short-term observing sessions. The additional equipment, a simple digital camera and home computer, are found in most homes of the twenty-first century.

The goal of a whole disc photography program is to obtain a sharp, clear, daily picture that shows the position of white light features on the disc. A photo should have a standard format, so a legitimate comparison can be made between photos from day to day. Proper orientation of a Sun photo will have celestial north at the top and east to the left, just as the Sun appears when on the meridian to a northern hemisphere observer, facing south. It is also appropriate to have markers within the photo indicating these directions accurately.

Some professional observatories use the daily value of P to adjust the orientation so that the solar N–S axis is vertical in the photo. Correct celestial orientation of a photo can be found by using the photographic alignment method with a crosshair or wire, as outlined in the previous chapter. An alternative method for defining directions would be to simply position the camera so that the four frame edges are parallel to N–S and E–W; you can do this by noting sunspot drift through the camera's viewing system. As a double check that the final photograph is properly oriented, make a comparison of the sunspot positions in the picture against the rough sketch you made for sunspot counting. It's easy to accidentally get an image flopped and have east for west or north for south.

A whole disc photo should also be produced as a photographic print in a standard 20 cm × 25 cm (8 × 10 in) format with the disc size being 18 cm. Although not set in stone, this is the dimension preferred by most professional astronomers; the Sun's disc is then large enough that detail to granulation size is visible. A digital file intended for on-screen viewing similarly could have a disc size of 18 cm, with a minimum resolution of 72 dpi. Pertinent information must be included with a photo, perhaps incorporated on the back side of a conventional photographic print or embedded in a digital file. Without this information the value of an observation is lost. A whole disc photo is best identified by the date and Universal Time it was obtained. Also included should be the name of the observer, information regarding the seeing conditions at the time of the photo, and technical data, such as telescope, filtration used, exposure, and recording media. See Figure 5.3 for examples of typical whole disc photographs.

Naturally, the most obvious data available from a whole disc photograph is the current status of the photosphere. What is rotating into view around the east limb, what is disappearing on the west? What is happening elsewhere? The heliographic coordinates of any feature may be quickly reduced from the observation. A comparison of the coordinates obtained over a period of time for regions of varying latitude will illustrate the differential rotation of the Sun. A series of daily whole disc photos also paints an interesting picture of solar activity. The sunspot zones will become apparent as spots are seen to develop in the region extending to about 35° north and south of the solar equator. When graphed against time, the latitude of the sunspots will display an interesting migration pattern, from the polar to equatorial regions. Occasionally, an observer can recognize another pattern of sunspot activity within specific regions of longitude; this

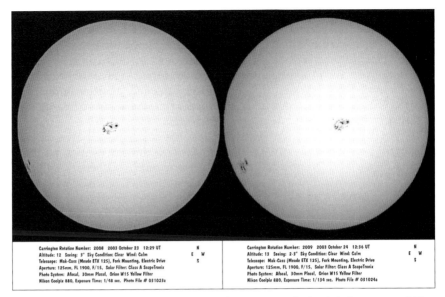

Carrington Rotation Number: 2008 2003 October 23 12:29 UT   N
Altitude: 12 Seeing: 3" Sky Condition: Clear Wind: Calm   E W
Telescope: Mak-Cass (Meade ETX 125), Fork Mounting, Electric Drive   S
Aperture: 125mm, FL 1900, F/15, Solar Filter: Class A ScopeTronix
Photo System: Afocal, 30mm Plossl, Orion W15 Yellow Filter
Nikon Coolpix 880, Exposure Time: 1/48 sec. Photo File # 031023a

Carrington Rotation Number: 2009 2003 October 24 12:36 UT   N
Altitude: 13 Seeing: 2-3" Sky Condition: Clear Wind: Calm   E W
Telescope: Mak-Cass (Meade ETX 125), Fork Mounting, Electric Drive   S
Aperture: 125mm, FL 1900, F/15, Solar Filter: Class A ScopeTronix
Photo System: Afocal, 30mm Plossl, Orion W15 Yellow Filter
Nikon Coolpix 880, Exposure Time: 1/134 sec. Photo File # 031024a

Figure 5.3. A pair of whole disc photos showing the movement of white light features in a 24-hour period. Howard Eskildsen.

could be explored via a whole disc photography program. Instructions for determining a heliographic coordinate through the use of a Stonyhurst Disc are found in Chapter 4 of this book.

A whole disc photo typically has a resolution of 2–3 arc seconds. Experiencing superb seeing conditions of 1 arc second or greater over the entire photosphere is rare; Earth's atmosphere is just not that forgiving. Nevertheless, even with mediocre seeing conditions, the evolution of many features is worthy of noting. The growth and decay of an umbra and penumbra; the twisting and rotating of a sunspot group; the development of a light bridge and small spots or pores that pop on the scene, only to disappear later will be evident in a whole disc picture. Perhaps even the seldom seen white light flare will burst forth during a photography session. Producing whole disc photographs is an excellent way of becoming familiar with the mechanics of the Sun. More importantly, the observer has documented a brief clip in the life of his subject. That moment can never be repeated, but it can be reviewed and analyzed any time in the future with the captured image.

After a photographer has perfected the necessary skill in this area of solar morphology, the next logical step is to create even more finely detailed photos of a specific region of activity. Close-up photos of sunspots, faculae, and the like requires patience, attention to detail, and just a bit 'o luck at the camera.

Active Region Photography

The goal of active region photography is to create detailed images of activity centers on the Sun that have a resolution of 1 arc second or greater. High-resolution photography of the Sun can be particularly challenging for the novice.

Highly technical skills are not necessarily required, but the atmosphere is constantly working against us. Seeing conditions and instrument defects are multiplied as an image of the Sun is enlarged to facilitate viewing finer detail.

The atmospheric conditions necessary for this work typically occur only during short periods of the day, meaning that "being in the right place, at the right time" is crucial. Granulation serves as an excellent indicator of seeing for the solar observer. When the granulation is clearly visible, then the resolution is 1 arc second or better. If the granulation is blurry, but can still be made out, then the seeing is on the order of 2–3 arc seconds. Anything worse is considered poor seeing, and photography would not be worthy of your efforts unless, of course, some feature is experiencing an important evolutionary change, putting the observer in a now or never situation.

While whole disc photography is ideal for the amateur capable of observing the Sun on nearly every clear day, those photographing active regions can observe only when they are able or if an unexpected event warrants special attention. Once an established routine is in place, a whole disc photo can be taken quickly. Active region photography, on the other hand, requires allocating a longer observing session, during which the amateur monitors the seeing conditions and photographs when the view is sharpest. If your desire is to tell the complete story of a happening, an hour or so daily at the telescope for the event's duration is standard.

Figure 5.4 is an example of a moderate resolution photo series showing evolution within AR0898 in approximately 1-day increments, beginning on the July 3 and ending on July 8 in 2006. A symmetrically shaped sunspot, AR0898 became a naked-eye beauty a few days before this series was started. Notice how the umbra took on an elongated shape immediately after the first day of observation. An encroachment on the umbra by the light bridge followed shortly thereafter and then continued with the breakup and decay of the umbral material. Image "f" is a bit misleading, due to the foreshortening of the sunspot as it approached the west limb of the Sun.

Although a complete well-timed photo series is impressive, individual snapshots are important, too. Sometimes we just can't observe for several days, and the project ends up containing sporadic or irregularly spaced photos. Often our seeing conditions prohibit a uniformity of quality to photos, and many pictures become unusable. The occasional high-quality photo does contain important information not always recorded by other observers. The story a single photo tells us is this: Here's what was happening on the Sun at this time. When a notable photo is combined with other similarly outstanding photos, the complete story of an event may be realized. Your singular picture could be the missing piece to a puzzle.

An unusual feature the solar observer can photograph is the phenomenon of polar faculae. This is a small, low contrast transient feature that appears beyond

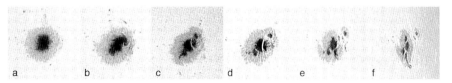

a   b   c   d   e   f

Figure 5.4. A series of images taken 24 h apart of AR0898, beginning with image "a" on July 3, 2006. Gema Araujo, except for "c," which is from Jamey Jenkins.

the sunspot regions. Observing with a green or blue supplementary filter enhances the tiny, almost dot-like pieces of faculae. An individual snapshot showing the location and shape of a polar facula is interesting, but again a series of photos capturing the birth and decay of this feature would be educational. The relatively short lifespan (in some cases only several minutes) of polar faculae means that a single observing session could possibly accomplish the project.

Similarly, monitoring the faculae that appear near the eastern limb in the sunspot zones could provide an opportunity to capture the early development of a sunspot group. A sunspot begins as a pore that forms near faculae, growing into an umbral spot that continues to develop a penumbra and eventually becomes a mature sunspot. Not all pores will go through this evolutionary pattern; many fade and decay before reaching the umbral spot phase. By maintaining a watch on a group of pores, in particular those associated with a large, bright facular region, the complete life cycle of a sunspot group over a period of several weeks or longer might be recorded. Observing a bright facular region that contains no pores could provide an opportunity to photograph the emergence and demise of individual pores (Figure 5.5).

An advanced project the amateur could pursue is the photography of detail in a sunspot's umbra. This is basically accomplished by increasing the normal photospheric exposure several fold until sufficient density has accumulated in the darkest region of the sunspot. Photospheric granulation and penumbral structure will unavoidably be overexposed and burned out, but granules, umbral dots, bright points, distinct detail within certain light bridges, and the core will become visible. The core of a sunspot is the area in an umbra having the greatest magnetic strength. Because this region will have peak density, it also has, according to the Stefan-Boltzmann law, the coolest temperature.

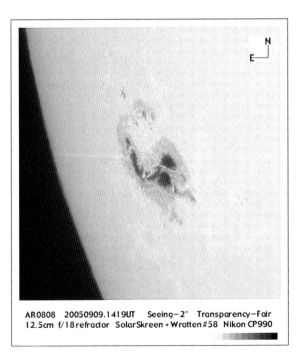

AR0808   20050909.1419UT   Seeing–2"   Transparency–Fair
12.5cm f/18 refractor   SolarSkreen + Wratten #58   Nikon CP990

Figure 5.5. Active region patrol photo. Jamey Jenkins.

Excellent seeing conditions are mandatory to create an effective deep umbral photo; otherwise the granular detail becomes smeared. Bracket the exposures to get it just right, because umbrae vary in brightness from one to another. An additional approach is to introduce an adjustable diaphragm at the first focus of the telescope to block out all light, except that of the umbra. The extraneous scattered light is detrimental to low-contrast umbral features. Upon going deep into an umbra, you'll find interesting structures ordinarily missed by conventional photographic procedures.

Another means for examining a sunspot umbra and the region surrounding it is the creation of an isophote contour map. An isophote is a grayscale or color diagram that silhouettes areas having the same or similar density. For a visual observer, recognizing the delicate contrasts within an umbra can be challenging. Using this technique of solar study, the observer can expect to identify a number of interesting features, including the inner and outer bright rings, weak light bridges, and the point of minimum intensity within the umbra, the core.

The generation of a modern isophote contour map from a high-resolution photo requires the use of image editing software on a computer. In days past, darkroom techniques would have required one to spend hours creating masks and films that separated out the different density levels in a negative. Today, the job is left to a point and click of the mouse to produce the density-separated files.

The Figure 5.6 images are typical of what can be learned from an isophote map of a photograph. This photo from April 5, 2001, is of the group AR9415 making its way around the east limb of the Sun. Areas in the map that have the same shade are regions of similar photographic opacity and consequently a similar solar temperature. Note that the leading spot (westerly) in both pairs have the greater density, marking the location of the umbral core in each. That is the location of the umbral core in those spots. Instructions for creating an isophote contour map will be found in a later chapter, entitled *Solar Photography*.

Like a whole disc photo an active region picture must be documented with the date and time (to the nearest minute) it was taken. Label the photo with the AR number of the region being photographed. The status of the seeing conditions and

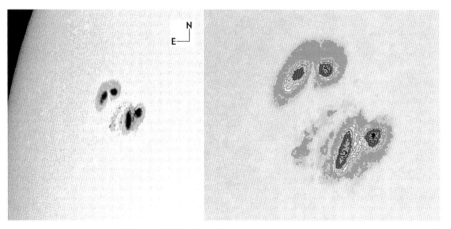

Figure 5.6. A conventional photo of AR9415 and an isophote contour map created to show areas of similar density and temperature from the left-hand image (see text). Jamey Jenkins.

transparency of the sky is to be recorded, as well as technical information concerning the telescope, filtration, exposure – and don't forget the observer's name. This data can be attached to the back of a conventional print or embedded in a digital image.

Viewing of the final on-screen image or print is convenient if the orientation is standardized with north up and east to the left. This matches a whole disc photo having the Sun as it appears in the sky on the meridian for a northern hemisphere observer facing south. It is customary to place pointers in the final picture indicating directions in the photo. You might also make a rough sketch at the telescope of the sunspot or other feature you're photographing, marking the celestial directions on it for quick reference when processing the image at the computer.

Reference

1. *How to Observe the Sun Safely*, L. Macdonald, Springer-Verlag, 2003

Chapter 6

Observing the Monochromatic Sun

Above the Photosphere

In Chapter 1, we used the analogy of a baseball to describe a number of layers that make up the interior of the Sun. The outer covering or skin of the ball was synonymous with the first of several more layers that constitute the solar atmosphere. This layer, called the photosphere, is seen in the integrated light of the solar continuum. The home of sunspots and other phenomena, it is also the source of the largest amount of photons from the Sun. The sheer brightness of the photosphere dominates the less harsh outer atmospheric layers, the chromosphere and the corona.

The region directly above the photosphere is the chromosphere (sphere of color), a less dense, nearly transparent gaseous sector about 2000-km thick having an average temperature of 10,000 K. Within this layer is an abundance of spectacular activity the amateur astronomer may study. Spicules, prominences, filaments, and flares are a few of the regularly observed chromospheric features. The presence of the chromosphere is confirmed during a total solar eclipse by the reddish-pink ring encircling the edge of the Sun at the time of totality. My first opportunity to scrutinize the chromosphere was during the February 1979 total eclipse from Brandon, Manitoba, in Canada. Wholly overwhelmed by the beautiful ring of color with its attending prominences, from then on I became captivated with solar astronomy. Many have compared the experience to their first sighting of the planet Saturn through a telescope. Having done both I can verify that like Saturn, spotting the chromosphere is indeed an awe-inspiring experience. Above the chromosphere and extending into space is the corona, the pearly white outer atmosphere of the Sun.

The corona is a mixture of colors, running the gamut from violet to red. The brightness of the corona is extremely weak when compared to the photosphere, on the order of a million times fainter. Other than during a solar eclipse, the only means of viewing it is with a sophisticated professional instrument called a coronagraph. An Earth-based coronagraph is a finicky instrument requiring the utmost cleanliness of its components, which includes an occulting disc to mask out the blinding light of the photosphere. Normally located high in the mountains to be above the scattering effects of Earth's atmosphere, professional coronal observations are best performed aboard spacecraft, or from satellites specifically

J.L. Jenkins, *The Sun and How to Observe It*, DOI 10.1007/978-0-387-09498-4_6,
© Springer Science+Business Media, LLC 2009

designed for solar work. Studies of the corona by an amateur astronomer are left to the several minutes of totality experienced during an eclipse.

As was said earlier, seeing the chromosphere is difficult because of the significant difference in brightness between it and the underlying photosphere. This is a lot like trying to see a burning match placed before the concentrated beam of a high-powered searchlight. The beam from the searchlight dwarfs the tiny flame unmercifully, just as does the photosphere when we attempt to see the weaker chromosphere. What is needed is a way to effectively turn off the photosphere. We can't turn off the light of the photosphere, but what we can do is to block its excess light and in that way isolate the light of the chromosphere.

The modern solar observer accomplishes this by viewing the Sun through a special filter. Excess light originating in the photosphere is effectively filtered out, leaving only the chromosphere to dominate the view. Such a perspective is called monochromatic, meaning that the view is of one color.

Emission and Absorption

If a ray of light from the Sun is spread into a rainbow using a dispersing device such as a prism or diffraction grating, then is inspected closely with the eye, a number of fine dark lines are seen to cross the spectrum. Joseph Fraunhofer noted this oddity in the early 1800s. Now called Fraunhofer or spectral lines, the lines and patterns they create serve as a compositional signature of the elements found in the photosphere. Spectral lines are the "Rosetta Stone" that allows astronomers to understand the chemical makeup of our Sun, the stars, and the distant galaxies so many millions of light years away.

In the laboratory, chemists discovered that if an element is heated until vaporized and emits light, and then viewed with a spectroscope, a set of lines unique to each element would appear and could be studied. What the chemists also found from this emitted light was that the Fraunhofer lines were displayed not as dark lines but as bright ones. Elements tested in this manner had bright lines that corresponded to an identical set of dark lines that could be found in the photosphere. Why were the lines dark one time and bright another? What causes the spectral lines in the first place?

Gustav Kirchhoff was a nineteenth-century physicist studying and cataloging the spectral lines of elements. Through his work came the laws that determine our present-day understanding of spectral analysis. Kirchhoff's laws state that: (1) A hot, dense, glowing body produces a continuous spectrum lacking spectral lines; (2) View a continuous spectrum through a cooler, transparent gas and dark lines called absorption lines appear, dependent on the energy level of the atoms present in the intervening gas; and (3) Hot, transparent gases emit the bright spectral lines, called emission lines, dependent on the energy level of the atoms.

The key to Kirchhoff's laws is found in the relative temperatures of the foreground and background bodies seen by the observer. If the foreground is cooler than the background, absorption lines result; when the foreground is hotter than the background, emission lines appear. The Sun produces a continuous spectrum without spectral lines below the photosphere because the gas there is dense and opaque. The photosphere produces dark spectral lines because it is cooler than the underlying dense layers. The chromosphere, particularly at the limb, appears bright because of a cooler background sky.

The absorption/emission spectral characteristics arise because atoms, which make up the elements, absorb or emit photons (light) at particular wavelengths. The hydrogen atom, the most abundant atom in the universe and in the Sun, is also the least complex, being composed of a single proton nucleus and a single orbiting electron. Electrons can have different orbital states, or what are referred to as energy levels. When absorbing photon energy, an electron jumps to a higher orbit or energy level, and when photon energy is released or emitted during excitation, an electron jumps to a lower energy level or orbit. These changes in energy level, the absorbing or emitting of photons, is what produces the Fraunhofer lines seen in a spectrum.

Hydrogen, in the visible wavelengths of light, produces several spectral lines that are called the Balmer series. Named for the schoolteacher, Johann Balmer, who recognized the pattern they form, the first of the Balmer lines located at 656.3 nm is designated, hydrogen alpha (Hα or H-alpha). The significance of the H-alpha line is that it is one of the brighter emission lines of the chromosphere. The reddish-pink color of the chromosphere is owed to the dominance of the H-alpha line and its location in the red part of the spectrum.

Another noteworthy line of emission in the chromosphere is located in the blue at 393.3 nm, the K-line of singly ionized calcium, referred to as the Calcium K-line (Ca-K). In the chromosphere, H-alpha and Ca-K are dense and opaque, but to the continuum light originating below the chromosphere, H-alpha and Ca-K are translucent. Consequently, the photosphere becomes feeble as we observe light only from those spectral lines. Using an appliance to pass only the specific wavelength of H-alpha or Ca-K light, the observer is able to see chromospheric features.

History of Chromospheric Observing

The specialized filter used by today's astronomers for monochromatic viewing of the Sun is relatively new on the scene. Prior to around 1930 the only method of visually seeing the chromosphere outside of a solar eclipse was through the clever manipulation of a spectroscope, or with a derivation of the spectroscope called a spectrohelioscope that synthetically creates a monochromatic view. The spectroscope is an optical device used for separating light into its component colors. The progress made in instruments from the early years of monochromatic observing is interesting and worth reviewing, if only for coming to an understanding of how we got to where we are today.

In the past, disturbances seen at the solar limb called prominences were visible only during those infrequent opportunities of a solar eclipse. Astronomers had difficulty determining if the prominences were phenomena of the Sun, the Moon, or Earth's atmosphere. One hypothesis was that they were clouds on the Moon, made visible by sunlight passing through them at the time of totality. Prominences remained a mystery to be studied fleetingly until the use of photography for astronomy in the mid 1800s. Initiated at an eclipse in July 1860 by Warren De la Rue, photography finally recorded the true essence of the prominences. Still, though, the chromosphere's limb features were only visible during the several minutes of totality.

At Guntoor, India, the solar eclipse of 1868 provided Jules Janssen of France the opportunity to examine prominences with a spectroscope attached to his telescope. From his observation he found that the spectral lines observed were in emission and quite bright. He also speculated that it might be possible to observe these lines in broad daylight. The following day he looked and again located the bright lines, finding that he could trace the shape of a prominence by shifting the position of the telescope and spectroscope with respect to the Sun.

Norman Lockyer, an English astronomer, weeks later duplicated the observation of Janssen, but with a delay in announcing their findings, both became credited with the discovery. Later that year, William Huggins, another English astronomer, managed to catch sight of a solar prominence through his spectroscope. Huggins found that by opening the slit on the spectroscope wider it was possible to see a greater expanse of the prominence. The spectroscope remained the workhorse of the solar astronomer studying limb features in monochromatic light for the next several decades (Figure 6.1).

A prominence spectroscope functions simply and is within the means of an amateur telescope maker to construct. The components consist of an adjustable entrance slit, a collimator, the dispersing element (in modern times usually a transmitting or reflective diffraction grating), and a viewing telescope. The entrance slit is positioned at the focus of the main telescope. The task of the collimator is to take the beam of light exiting the slit and make it parallel, so as to flood the dispersing element with light. The diffraction grating is at the heart of a spectroscope. Ruling fine grooves on a low expansion substrate, such as glass, creates a diffraction grating. Light is dispersed as it is transmitted or reflected from the grating. A grating's refractive result is similar to that of the prisms used by earlier observers, but with greater efficiency. The purpose of the viewing telescope is to provide a means of observing a magnified view of the spectrum.

Observing a solar prominence is accomplished by focusing the limb of the Sun adjacent to the entrance slit of the spectroscope, with the H-alpha line centered in the viewing telescope's eyepiece. Carefully open the slit and inspect the field to see if any prominences are visible. If none is, then reposition the spectroscope to another area of the limb. There is a limit to how far an entrance slit may be opened because eventually the background becomes excessively bright and the prominence difficult to observe. Depending on the mounting arrangement for the spectroscope, this may be the cumbersome part of prominence studies. Much time and energy is consumed manipulating the position of the spectroscope relative to the solar limb.

In 1891, George Ellery Hale suggested a method of introducing moving slits to the spectroscope that allowed photographing the Sun in monochromatic light. His

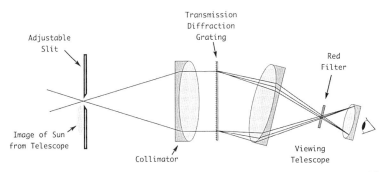

Figure 6.1. The optical layout for a prominence spectroscope using a transmission diffraction grating. The red absorption filter blocks wavelengths outside the region of the H-alpha line.

instrument, called a spectroheliograph (SHG), had been proposed a number of years earlier, but Hale independently re-invented the technique and was the first to actually put it to use. The SHG has an advantage over modern narrow band filters of permitting an observation at any wavelength in the visual spectrum. Because the photographic materials on hand were primarily blue sensitive, Hale selected the blue line of Ca-K when capturing prominences. Besides establishing the world's largest observatories several times over, George Hale made numerous discoveries in the field of solar physics. Attributed to Hale are the discovery of the magnetic fields of a sunspot as well as cloud-like regions of calcium flocculi (Figure 6.2).

The SHG uses two slits to form a monochromatic image of the Sun. The entrance slit, collimator, grating, and viewing telescope remain essentially the same as in a spectroscope. The difference between the SHG and spectroscope is the replacement of the viewing eyepiece with an exit slit near the detector, which isolates an individual spectral line for observation. By scanning the two slits in unison across the solar disc, an image of the Sun in the selected wavelength is assembled, one slit width at a time. Replace the detector (film or electronic) with an eyepiece, and a visual counterpart, the spectrohelioscope, is created. A number of craftsmen since Hale's time have constructed several variations on this instrument. The difference between most designs involves how to synthesize the final images. Oscillating slits, rotating prisms, and movable mirrors are a few of the possibilities telescope makers have explored. A spectrohelioscope or graph tends to be a bulky

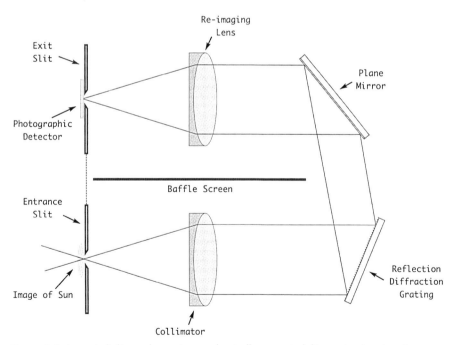

Figure 6.2. A spectroheliograph uses two mechanically connected slits moving in unison to scan a wide field view onto a photographic detector. Replace the detector with an eyepiece, increase the speed of the oscillating slits until the eye perceives a composite image, and a spectrohelioscope is created.

instrument, not very portable, and is best utilized when in a fixed position and fed by either a heliostat or a coelostat.

A heliostat consists of one or more mirrors that reflect light from the Sun to a stationary telescope. The mirrors must be motor-driven or the Sun will continually drift out of the field of view. Rotation of the field with a heliostat-fed telescope is a drawback to this system. If two mirrors are mounted on a motor driven coelostat, the Sun will remain stationary and not rotate about the optical axis of the telescope. The coelostat is a complex arrangement having one mirror positioned at the latitude of the observing station; the second mirror directs sunlight to the telescope.

A spectrohelioscope is a powerful instrument capable of performing high quality observations. Studies in selected wavelengths of light, the magnetic properties of sunspots, and velocity investigations of erupting features are possible.

In the early 1930s a new innovation was suggested by the French astronomer, Bernard Lyot (also the inventor of the coronagraph) and further developed by Jack Evans. The device would allow direct observation of the Sun in monochromatic light. Not requiring the dispersion of sunlight as in the spectroscope, a Lyot filter works on the principle of light interference known as birefringence to isolate a band of monochromatic light. Birefringence, the double refraction or splitting of light, takes place when a beam of light encounters special optical glasses, notably calcite or quartz crystals. The two sequential beams, polarized at right angles to each other, pass through the crystal at differing velocities. Depending on the thickness of the crystal, it becomes possible to have one beam gain half of a wavelength on the other, that is, be 180° out of phase and interfering destructively. By combining a set of crystals of varying thicknesses, a birefringence filter passing a narrow bandwidth of light at any wavelength may be constructed.[1]

The new Lyot filter permitted direct observation of H-alpha light in a compact package, something never experienced before. Early filters had a bandwidth greater than 1 Å and allowed observing only prominences, but as narrower bandwidths became available, surface detail was perceivable. The high-end professional Lyot filter today is an extremely expensive product, beyond the means of most amateur astronomers. A Lyot filter is customarily tunable through a wide range of wavelengths by rotating the elements of the filter. The crystals are rare and construction is extremely precise. In the amateur ranks a few names from of the past standout as having accepted the challenge of building a birefringence filter, such as Henry Paul and Walter Semerau (Figure 6.3).

A colored glass filter of the Wratten type works by absorption; material within the glass literally absorbs certain colors of light while the color of choice passes

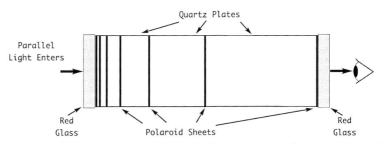

Figure 6.3. A birefringent monochromator or Lyot filter uses exotic glass in successive double thicknesses to reinforce or interfere with light rays.

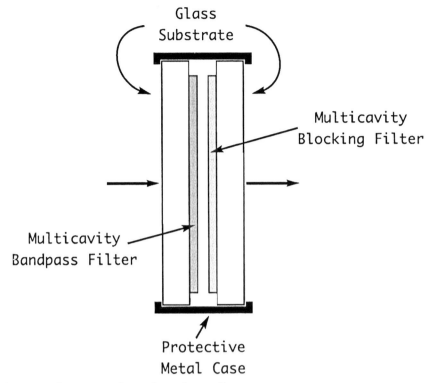

Figure 6.4. Construction of a simple interference filter.

right on though. This type of filter, however, has weak control over the cut-off of the wavelength of light passing through it. In other words, a green filter may include a small amount of blue and yellow light. Another filter similar in appearance to a Wratten but operating completely differently and capable of transmitting narrow bandwidths is called an interference filter.

During the late 1800s, Charles Fabry and Alfred Perot investigated the possibility of an interferometer based on multiple reflections between two closely spaced, partially silvered mirrors. Interferometry is the superimposing of multiple light waves created by the interferometer for the purpose of studying the differing characteristics between an incoming wave of light and an outgoing wave. Modern interference filters, briefly discussed in Chapter 3, are modeled after the Fabry-Perot interferometer.

An interference type of filter is made possible because of the development of a thin film optical coating technology during the 1930s. Fabricated by depositing materials with a thickness comparable to a wavelength of light on precisely made substrates, a thin film filter is designed to exclude much of the off-band light that an absorption filter transmits. The transmission bandwidth of the modern interference filter has been reduced to 0.1 Å. When one of these specialized filters is applied to solar observing, a breathtaking view similar to that of the Lyot filter is obtained, but in an even more compact package and at a reduced cost.

In the basic interference filter, a separation exists between the reflecting surfaces comprised of a thin dielectric material called a spacer. The reflecting layers themselves are made of several film deposits of high and low index materials, sometimes zinc sulfide or cryolite salts. These successive layers deposited on the substrate are known as a stack. Two stacks with the intervening spacer amount to what is called a single cavity bandpass filter. The number of layers found in a stack and the thickness of the spacer is modified to suit the desired bandwidth and wavelength output. This elementary filter design does not limit all the unwanted wavelengths passing through a filter. In order to remove the excess wavelengths, additional layers are incorporated, or a package of several filters is sandwiched, creating a multi-cavity filter.

An interference filter is sensitive to the angle of the light passing through it. The thicknesses of its components, as well as the tilt of the filter relative to an incoming beam of light, are critical to the filter's output. Steep angles off an optical axis increase the distance light must travel through a filter, effectively shifting a filter's transmittance to a lower wavelength and widening the bandpass. The amount of wavelength shift experienced is dependent upon the incident angle and the refractive index of the filter. In the ideal situation, parallel light rays are chosen to pass through an interference filter. The expansion and contraction of the substrates and coatings from variations in temperature also alters the specifications of an interference filter, necessitating in some cases a temperature-controlling oven to house the filter. Since linear wavelength shift will occur with changes to the ambient temperature, a filter is constructed to have a specific operating temperature.

The thin film interference filter has opened the floodgates for the amateur astronomer wishing to explore the Sun at a specific wavelength. The cost of a filter is sometimes no more than that of a pair of modern eyepieces and within the pocketbook of many observers. However, don't be misled; the higher-end filters are very expensive accessories. Regardless, the view of the Sun through one of these devices is awe-inspiring and truly breathtaking, reminiscent of an amateur's first view of Saturn through a telescope.

Before the discussion continues, it would be worthwhile to review a number of frequently used filter and monochromatic terms.

Filter Terminology

Bandpass or bandwidth – the extent or band of wavelengths transmitted by a filter.

Blocking – the amount of light attenuation at wavelengths outside the bandpass of a filter.

Center Wavelength (CWL) – the wavelength found at the midpoint of the full-width half-maximum.

Double Stacking – a method of narrowing the bandwidth of an etalon by the addition of a second etalon.

Energy rejection filter (ERF) – a pre-filter that is placed over the opening of the telescope for the purpose of absorbing or reflecting UV/IR light and reducing the heat load on the interference filter.

Etalon – an optical filter that operates by the multiple-beam interference of light, reflected and transmitted by a pair of parallel flat reflecting plates. Principle based on the Fabry-Perot Interferometer.

Field angle – the angle of outside light rays entering a telescope. One example is illustrated by the angular size of the Sun as it appears in the sky.

Full-width half-maximum (FWHM) – the measured width of the bandpass, in nanometers or angstroms, at one-half of the maximum transmission.

Instrument angle – the angle of light rays converging to a focus in a telescope.

Interference filter – an optical appliance with several layers of evaporated coatings on a substrate, whose spectral transmission characteristics are the result of the interference of light rather than absorption.

Monochromator – any device that produces a narrow band of monochromatic light.

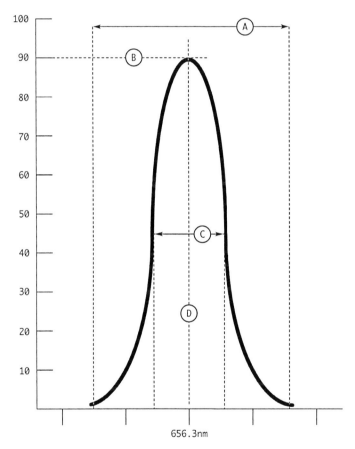

A = Bandwidth
B = Peak Transmission
C = Full-Width Half-Maximum (FWHM)
D = Center Wavelength (CWL)

Figure 6.5. The transmission profile is used to describe filter characteristics. It graphically illustrates the bandwidth, peak transmission, full-width half-maximum, and center wavelength. Because each tick mark at the foot of the profile represents 0.1 nm: A = 0.3 nm, B = 90%, C = approx. 0.1 nm, and D = 656.3 nm, the H-alpha line.

Nanometer – the nanometer (nm) is a unit measurement of wavelengths of electromagnetic radiation (light). One nanometer is equal to one billionth of a meter (1×10^{-9} m). Another frequently used light wave measurement is the angstrom (Å), 1×10^{-10} m, or .1 nm.

Normal incidence – light rays that follow a normal or parallel path.

Occulting disc – usually a cone-shaped polished metal component for blocking the light of the photosphere in a prominence telescope.

Oven – an electrically controlled device for regulating the temperature of an interference filter.

Peak transmission – the maximum percentage of transmission found within the bandwidth.

Telecentric lens – a supplementary lens system intended to create normal incidence light rays from the converging light rays of a telescope.

Selection of a Telescope

In an earlier chapter the choice of a telescope intended for white light observing was discussed. The conclusion was that any telescope could be used in that venue if it sufficiently limits the radiation harmful to the human eye. Some designs are naturally more "user-friendly" than others, and lend themselves to optimal observations of the Sun. A similar situation exists for a monochromatic observer using a modern interference filter on his or her telescope; some telescopes are perfectly suited, while others are less than ideal.

It pays to take a look at the requirements for use of an interference filter when deciding on a telescope intended for monochromatic observing. An interference filter delivers performance when normal incidence or parallel light is passed through it and when the filter resides in an environment having a stable operating temperature. A filter system can take up several inches (1 in = 25.4 mm) of back focus on a telescope and all require that a pre-filter, called an energy rejection filter (ERF), is placed externally over the telescope objective as pictured in Figure 2.1. An ERF is an optical grade filter sometimes made from Schott RG red glass, some other absorption glass with similar transmission properties, or a dielectric coated "hot mirror." The important criteria are that the ERF prevents IR/UV light from entering the telescope assembly and that it has a polished surface accuracy of at least a quarter of a wave.

Field angle is the path taken by rays of light entering a telescope from an object that is suspended in the sky. It is interchangeable with the more commonly used term, field of view. The Sun presents an angular displacement in our sky of approximately 32 arc minutes, meaning that the light at the solar limb travels at an angle of about 16 arc minutes relative to the center of the Sun. Some rays reach the telescope at normal incidence, in parallel bundles, while those farther from the Sun's center are at an ever-increasing angle. This is of consequence to the solar observer with a large diameter etalon over the telescope objective because some rays contributing to the image will not be incident. Those rays on the angle will have a longer path through the filter, resulting in a widening of the bandpass.

Instrument angles are those generated by the optics of a telescope. A diagram of a telescope's optical layout depicts the route taken by light near the edge of the objective. These rays converge on a cone-shaped path to a point of virtual focus called the first or primary focus. Light rays contributing to the virtual image that are on the optical axis are of normal incidence; those closer to the edge of an optic have a steeper path to reaching focus. As with the field angle, instrument angle can result in a lengthened path through an internally placed filter, again contributing to the widening of a filter's bandpass and a shifting of the center wavelength.

A Newtonian reflecting telescope will be limited because of the minimal back focusing range available on most commercial instruments. To resolve the issue, the point of final focus must be moved a greater distance outside the tube of a Newtonian by decreasing the distance between the primary and secondary (diagonal) mirrors. The disadvantage here is that to eliminate vignetting of the image, a larger secondary mirror is required. This introduces a larger obstruction before the primary mirror and a decrease in contrast to the final image. For an astrophotographer, a mirror presents the advantage of bringing all colors to a common focus, but in monochromatic observing this is for naught. A monochromatic observation is accomplished by using only the tiniest portion of the visual spectrum, a single color brought to focus. This is a task that could be performed by a simple plano-convex lens. The Newtonian telescope, while adaptable to chromospheric observations, does not present a distinct advantage with this type of observing.

A catadioptric telescope often uses a short-focus concave primary mirror and matching convex amplifying mirror to obtain a long focal length in a compact package. The amplifying magnification of the secondary mirror is often three or more times, affecting negatively the instrument angles and resulting in a "sweetspot" of on-band transmission surrounded by an area of increasingly off-band light. Schmidt-Cassegrain and Maksutov telescopes, while portable and fine instruments for white light solar observing, are not the optimal choice for the monochromatic solar observer.

The refractor, with its straight through optical path, a lack of central obstruction, and typically long back focus make it the preferred instrument for monochromatic observing. To combat the effects of field and instrument angles, some manufacturers recommend stopping a telescope down to a focal ratio of f/30 or greater. This provides the situation of almost incident light, one in which a negligible shifting and widening of the bandpass occurs, and one that most observers choose to ignore. A refractor permits stopping down a telescope on-axis so the normal incident light rays are used. The refractor avoids odd instrument angles as are encountered with a catadioptric or Newtonian that has been stopped down off-axis. However, a disadvantage of stopping down any telescope to f/30 will be the loss of resolution. For instance, a 125 mm f/10 telescope stopped to f/30 has an aperture of only 42 mm. Imagine being confined to exploring the Moon and planets with a telescope of only a 42 mm aperture!

One way around the instrument angle verses resolving power issue has been through the use of an add-on lens system that forces the converging cone of light into a nearly parallel beam before admitting it to an interference filter. Such a system is called a telecentric lens. The addition of a telecentric allows for a full aperture ERF. Like a Barlow lens, a telecentric will magnify the focal length of a telescope, but without the Barlow's diverging of light rays. Unfortunately, a Barlow lens generally doesn't work well with an interference filter because it magnifies the field angles. The telecentric system should be fitted for the specifications of the telescope in use. A custom made unit is an expensive accessory and a few high-end telescope manufacturers supply a telecentric as an add-on for their instruments.

Many experienced solar observers have, as an alternative to the telecentric, used the Powermate[TM] amplifying lens from Televue. This unique product produces instrument angles that are uniform across the image plane, not divergent, as with a Barlow, and does not require customizing for a specific telescope. The Powermate is also available in several magnification factors that increase the focal length substantially for a short- to medium-focus telescope. Always confer with the manufacturer of the monochromatic filter to find the adaptability of any add-on accessories, such as a telecentric or an amplifying lens.

Filtering Systems for the Telescope

There are two basic approaches for adding a narrow band solar filter to an amateur's telescope. One is front loading and uses a large diameter etalon over the entrance to the telescope, a mode similar to the placement of a white light objective filter. The second is called end loading. An end-loaded etalon is smaller than most front-loaded filters and is positioned inside the light path of a telescope, in front of the eyepiece. Each style performs well, having distinct advantages and disadvantages. In the following pages we will discuss applications of each filter design. The amateur should select a monochromatic filter that best suits his or her observing needs. Do you have a particular interest in studying disc detail (filaments, flares, plage, etc.) or would you be content observing only prominences? This is an important question to ask, because the answer will determine the characteristics and the cost of the chosen filter.

To observe prominences at the limb in H-alpha requires a filter with a bandpass no wider than 10-angstoms (1 nm). A wider bandpass than this and the lack of contrast between the prominence and the background sky becomes a problem. An additional internal occulting disc or cone to hide the photosphere is necessary; the disc of the Sun will be uncomfortably bright for the observer without obstructing it. The H-alpha coronagraph by Baader Planetarium is a commercially manufactured example of an add-on device designed for prominence-only viewing. Although it functions beautifully, you should be aware that a prominence viewer like the Baader unit is not so compact, and adds about 200 mm to the length of the telescope.[2] Such an accessory can cause an out-of-balance situation or make for an unusual position when observing at the zenith. This is an end-loading type filter that because of its wide bandpass puts less demand on instrument angles than a narrow band model. Water vapor and dust in the atmosphere, however, are particularly detrimental when observing through one of these filters. Contaminates add significantly to scattered light and a loss of contrast, but on a calm, sunny day with a deep-blue transparent sky, solar prominences stand out remarkably well.

The narrower a filter's bandpass, the more the H-alpha line is isolated from the light of the solar continuum and the greater is the contrast of the resulting image. A narrow bandpass filter, however, is costly to construct because it is more complex! The 10-angstrom filter reveals only emission features against a dark background sky. A narrower bandpass filter of 1.5-angstroms (0.15 nm) will show prominences well, some Doppler effects, the bright flare, and, when sky conditions are favorable, the darker filaments. This H-alpha filter will not require an occulting cone if its transmission characteristics limit the brightness of the solar disc. An energy rejection filter is a necessary precaution, guarding both the 10 and 1.5-angstrom filters from deteriorating UV/IR radiations and excessive heat.

Each of the above filters must also have a sharp cut-off in transmission that blocks any light outside the bandpass from reaching the eye. A filter having a bandpass of 1.5 Å or greater can be used in optical systems operating near f/20, but because the light is converging in a steep cone, a widening of the bandpass and a shift of the CWL will occur. This author has used a 10 and 1.5-angstrom H-alpha filter effectively with an f/18 telescope for prominence studies; a short focus telescope is not recommended for either of these accessories. Neither filter requires temperature control, but both function better if the surrounding air temperature is close to the filter operating temperature. The 1.5-angstrom filter must have a tilting mechanism to fine-tune the CWL. The filter will need to warm up when used with an ERF and not an occulting cone, causing it to significantly shift off band. The tuner will allow bringing it back on band while permitting the observation of Doppler effects on prominences and filaments.

A filter having a bandpass less than 1 Å (0.1 nm) is called "sub-angstrom." This is a precisely made appliance that passes light of only the thinnest slice from the solar spectrum. While a prominence filter costs in the hundreds of dollars, a H-alpha sub-angstrom filter is priced around several thousand dollars and upward. The sub-angstrom filter also requires an energy rejection pre-filter, near incident light rays, and usually an electrically controlled temperature oven and/or a tilting device for fine-tuning its transmission characteristics. Sounds like a lot of additional work when setting up a telescope for solar observing, and it is. But the chromosphere viewed through one of these filters is a truly awe-inspiring experience. A sub-angstrom filter permits clear observing of filamentary structure, flare phenomena, prominences, and all the associated chromospheric activity.

End-Loading H-alpha Filters

The H-alpha filter that attaches to a telescope near the tail of the optical path is called end loading. This filter's housing is constructed with endplates that accept standard threaded adapters for attaching it to the telescope and eyepiece. Both the sub-angstrom and wider bandpass filters are available as this type of filter.

At the time of this writing two manufacturers of the 1.5-angstrom H-alpha filter are producing ready-to-use units. Lumicon and Thousand Oaks Optical sell a package that includes the necessary over the objective ERF, plus an end-loading H-alpha filter mounted in a tunable holder. The H-alpha filter inserts into the telescope focuser, accepting an eyepiece or a camera adapter of standard barrel diameter. The owner must supply an f/20 or greater focal ratio telescopic assembly. These are very basic units that allow the observation of prominences and minimal surface detail. They are a great introduction to monochromatic observing.

Several companies manufacture sub-angstrom filters of the end-loading type. Noteworthy among the list is Daystar Filters LLC that began a line of compact off-the-shelf filters in the latter part of last century. The near-incident beam required by a filter may be obtained by using a small diameter ERF over the telescope objective, effectively masking to an f/30 aperture, or by inserting into the optical

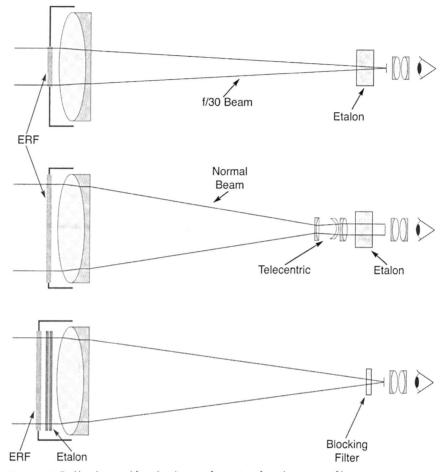

Figure 6.6. End-loading and front-loading configurations for sub-angstrom filters.

train a telecentric device, as outlined earlier. Daystar also markets a tilting, unheated model filter that provides maximum portability.

A sub-angstrom filter is manufactured with a normal operating temperature between 90°F and 150°F. The task of the heating unit then is to maintain the proper operating temperature. A warm-up period for an oven has been reported to be anywhere from several to many minutes; it all depends on the filter's ambient temperature. Decreasing or increasing the oven temperature a specific amount will regulate the transmission characteristics of the filter by an angstrom on either side of the desired CWL. When an observer intentionally causes the filter to shift off-band this is called viewing in the wings (i.e., blue-wing or red-wing of H-alpha). Observing in the wings allows the scrutinizing of a lower region of the chromosphere, or perhaps a Doppler-shifted feature that would otherwise be invisible on band. This happens because as we observe further from the CWL we are seeing light originating more so from the solar continuum, that is, light from the photosphere. Doppler-shifted features are moving at a high velocity, and their light waves become stretched or compressed, shifting the Fraunhofer lines slightly from their normal position in the spectrum. Some models of sub-angstrom filters utilize

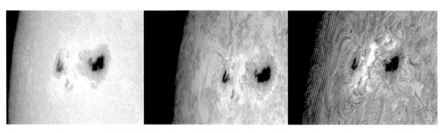

Figure 6.7. Images of AR8970 from April 22, 2000; from *left* to *right*, an adjustment of 2.5 and 1.0 Å towards the blue wing of H-alpha. The *right hand image* is on the CWL. Gordon Garcia.

a cooling fan in the event the ambient air temperature becomes greater than the operating temperature of a filter (Figure 6.7).

Mated with a telecentric lens and a full aperture ERF, an end-loaded sub-angstrom filter is ideal for realizing the full resolution of a solar telescope. A complete range of magnifications is available with the change of an eyepiece, allowing especially fine prominence or fibril detail to be visible through the filter. Unfortunately, a disadvantage to an electronically heated filter is the time factor when shifting a filter off-band. It can take many minutes to move the CWL several angstroms if comparing a feature's appearance on-band and in the wings. Most Doppler-shifted events occur rapidly and don't wait for a heated filter to catch up. To work around this situation, the housing of an end-loading sub-angstrom filter can be rigged to tilt, thereby giving immediate control of the CWL to the observer. More than one veteran observer has expressed a desire that the manufacturers of end-loading filters produce a version of the oven-heated filter with an additional internal tilting mechanism that permits a rapid shifting of a filter's CWL.

Front-Loading H-alpha Filters

A front-loading system sidesteps the difficulties of instrument angle by placing an etalon at the entrance to the telescope. The observer is able with this system to utilize various focal length telescopes without relying on external telecentric systems.

In the latter half of the 1990s, the firm Coronado Technology Group began manufacturing sub-angstrom H-alpha filters of the front-loading variety for amateur astronomers. With the Coronado design, both the ERF and etalon are located in front of the telescope objective. The etalon is stable over a wide temperature range, due to the unique spacing post between its components. All etalons transmit "side-bands," pieces of the spectrum other than that of the CWL, that must be removed with a so-called blocking filter. In a Coronado filter the blocking component is located at the eyepiece end of the telescope, an end-loading filter has the blocking filter incorporated within the filter pack of the monochromatic filter. A Coronado filter does not require electrical power for a heating unit, making it a very portable telescope accessory.[2]

The front-loading filter design has been very popular with amateur astronomers since its introduction. The H-alpha filter set with a matching blocking filter is available in several apertures having a bandpass of <0.7 Å and is easily adaptable to various telescopes with its corresponding mounting plate. This is a unit that can rapidly tune off-band by tilting the front etalon via a thumbscrew on the filter's mounting plate, a distinct advantage for an observer viewing a Doppler-shifted event, such as the disparition brusque of a filament. Possessing a telescope with a focal length less than an arm's length makes this adjustment easier. A longer tube may necessitate some sort of extension arm to permit convenient adjusting of the filter while simultaneously viewing through an eyepiece. Perhaps some ingenious solar observer will eventually assemble a motor drive arrangement that will adjust the tuning knob remotely.

To narrow the bandpass of a front-loaded filter to <0.5 Å, a technique called "double stacking" has been devised in which a second etalon is mounted in conjunction with the first. A narrower bandpass means increased contrast of a feature. There are several approaches to this with regards to how and where the second etalon is mounted, that is, whether internally or externally. The most practiced technique is to mount it externally, mated with the original etalon; each is then tuned separately to facilitate a total narrower band transmission. The important consideration here is that the second etalon should have certain transmission characteristics relative to the first, insuring a correctly functioning set. Double stacking becomes effective when each filter's transmission profile clips the other in such a way as to narrow the combined transmission of the pair (Figure 6.8).

In recent years Coronado has added to their line of dedicated solar telescopes an inexpensive instrument named the PST™ (Personal Solar Telescope) that has introduced monochromatic observing to many amateur astronomers. A self-contained 40 mm f/10 aperture refractor, the H-alpha model has a bandpass of <1.0 Å. It is difficult to attend any astro-event and not see at least a handful of these solar telescopes on display by their owners. The view through these, while not as contrasty as with a narrower bandpass filter, is still impressive, allowing prominences and disc detail to be visible. Nearly all PSTs seem to exhibit a "sweet-spot" near the center of the field of view where H-alpha detail is superior. Sweeping across the solar disc with one of these telescopes makes it possible to obtain a wonderful impression of the H-alpha Sun. This is a minor drawback for the visual observer and should not deter the observer from obtaining one of these telescopes.

Figure 6.8. Coronado front-loading 60 mm H-alpha filter set mounted on a Vixen 102ED telescope. Steve Rismiller.

A front-loading filter has the advantages of convenience and extreme portability. It is designed primarily for whole-disc viewing of the Sun, being limited only by its own aperture. An end-loading filter can allow maximum resolution; this sometimes necessitates the use of add-on telecentric lenses. Both designs provide an excellent view of the monochromatic Sun and its ever-changing features. When debating about which type of filter to obtain for your observing choices, be sure to try out various filters on other observer's telescopes. Talk with friends and solar observers, checking how a product met their needs and expectations. Such an expensive investment requires some thorough investigation, so that in the end you make a purchase that is right for you.

Filters for Ca-K Observing

The light of H-alpha shows a profusion of activity in the chromosphere, but located in the violet portion of the spectrum at 396.9 nm and 393.3 nm are spectral lines of additional interest to the solar observer. This light is from the emission of the H and K lines of singly ionized calcium. Compared to the H-alpha line, the H and K lines are broader and thicker in appearance. This means that to isolate a monochromatic view in the light of calcium, a filter does not need to have a particularly narrow bandwidth. Where a sub-angstrom filter will be necessary for good contrast in H-alpha, a filter having a bandwidth of 2–10 Å is sufficient for Ca-H or K observations.

What do we see in the light of calcium? A lower, cooler region of the chromosphere than in H-alpha is visible. The web-like chromospheric network assumes a bright-on-dark pattern, prominences appear blue, dark filaments are seen against the disc, and plages assume a bright cloudlike form (see Chapter 7). Unfortunately, observing at this end of the spectrum is frequently left to photographers. The human eye as it ages gradually loses sensitivity to violet light, making it difficult for a mature adult observer to visually see much of anything. One certain way around this is to connect to the telescope a video monitor that has a sensitivity to violet light and watch the on-screen image. Most amateur observations are made in the K-line, which ironically is the more difficult of the two to see, since it is deeper into the violet. If you find that you are unable to see in the K-line, a filter is available for H-line observing that may permit seeing some violet features. The H-line is located a bit closer toward the region where the eye becomes sensitive to light. Keep in mind, though, that either filter's suitability is totally dependent on the observer's eyes.

Coronado manufactures a dedicated Ca-K telescope with a 70 mm aperture having a maximum bandpass of 2.2 Å; it also produces a Ca-K version of the PST with a bandpass of 2.2 Å. An end-loading filter for Ca-K utilizes a telescope with a focal ratio of f/20 or greater, and to minimize heat buildup at the filter, a maximum aperture of 80 mm is suggested. High-end filters are normally temperature controlled via an oven, just like H-alpha products. We mentioned in Chapter 3 the Baader Calcium K-Line filter. This filter has a relatively wide 8 nm bandwidth centered on 395 nm to allow viewing of notable Ca-K features. The 80-angstrom bandwidth is not as efficient as one of the above narrow band filters, allowing more of the solar continuum to pass, but it does provide an interesting view. Use of this particular filter is recommended for photography only, because of the potential for high UV exposure to the observer.

Observing Tips and Accessories

Solar observing is performed in conditions very different from that of the evening star gazer. For instance, the light of the Sun produces a situation where a reflection from our eye creates a "ghost image" in the eyepiece. Exposed skin is readily sunburned from prolonged exposure to our subject. This will never happen when observing the Moon at night. Any experienced observer can tell you that a monochromatic filter can be a rather finicky device requiring special attention to cull from it all the possible information. To assist the observer a few homemade accessories make this level of finesse possible.

Amateur astronomers are clever innovators. It is difficult to pursue this hobby and not be tempted to tinker at least occasionally as an ATM (amateur telescope maker). Below are several observing tips for the solar astronomer plus discussion of a few custom accessories that have proven beneficial to other amateurs.

Enhancing Disc Contrast

To facilitate observing prominences and disc detail with his H-alpha setup, solar observer Greg Piepol has made use of an adjustable iris diaphragm over the telescope's ERF. The iris is a large-scale version similar to one located within the lens housing of a digital camera. Curved blades open and close to control the amount of light admitted to the camera, or in this case the telescope. According to Piepol, "With it (the iris) set at full open, prominences are at their brightest while disc detail is slightly washed out. For disc detail, I decrease the opening to about 95 mm. This darkens the view and allows me to see more of subtle features." Piepol's diaphragm is available from Edmund Scientific. This particular model is adjustable from a closed setting of 6 mm to fully open at 120 mm. Velcro is used to secure the iris to the lens cell, making an attractive and useful accessory for visual or photographic observing (Figure 6.9).

Figure 6.9. Adjustable iris diaphragm. Greg Piepol.

Other useful accessories for observing the H-alpha Sun are supplementary polarizing filters. Two polarizers are inserted after the H-alpha filter pack, usually screwed to the barrel of an eyepiece, and are adjusted by rotating one of them until a suitable transmission is obtained. The effect would be similar to the above iris in dimming the solar disc; some observers using these devices have also reported an increase in the contrast of features.

Sun Shades

Even a white light observer can benefit by being shaded from direct sunlight. Solar observing in a shadow reduces eyestrain, the potential of ghost images in the eyepiece, body fatigue, and an opportunity to get sunburned. When viewed through some telescopes, the Sun can be relatively dim and features difficult to see. One instance would be when attempting to spot features through a Ca-K filter. You hardly think of having to dark-adapt your eyes when observing the Sun, but there exist times when the technique is practical.

A telescope sunshade can be made from thin (1/8 in) wood, aluminum plate, or even stiff cardboard. Typically a circular shape is cut from the material about 300% larger in diameter than the telescope tube. Another hole is cut in the center of this disc of a size to slip over the skyward end of the tubing. Depending on the arrangement, the ERF or white light objective filter may be used to hold the shade in place. Paint the side facing the Sun white to reflect heat and the side facing the observer flat black. With a refractor or Cassegrain-type telescope the shade alternatively can be attached to the tailpiece of the tube assembly; this configuration does not stop light from reaching a solar finder or other piggy-backed telescope on the main instrument.

Not every observer is satisfied with a tube-mounted shade. Amateur solar observer Jerry Fryer once compared his tube-mounted shade to a sail on a boat in a breeze, each puff of the wind inducing vibrations in the eyepiece. What works well for Fryer and as illustrated with the Hess' heliostat in Figure 6.10 is a shade, mounted on a stand separate from the telescope. Fryer has constructed his from an inexpensive floor model lamp stand. It is similar to an expensive microphone stand, with an adjustable arm used to hold the shade. Such a design can be positioned as needed and does not contribute to the vibration causing "sail effects."

Some serious observers use a cloth, a hood, or just a large bath towel draped over their head and eyepiece to provide a portable darkened environment. Hood-like coverings of this type used on a portable spectrohelioscope can function superbly. A dark, tight-weave towel once served this author for viewing the Ca-K Sun and again to see its image on a laptop computer in broad daylight. Ideally, the hood or cloth should be black on the inside and white on the outside to help reflect heat from the observer, the ultimate goal being to exclude all ambient sunlight.

Extended exposure to sunlight (UVA/UVB radiations) can be dangerous to a solar observer's health. That is why sunscreen lotions are important for the body. During extended observing sessions use a product with a suitable sun protection factor (SPF). The arms, face, and back of the neck seem to be particularly vulnerable. When solar observing, a long-sleeved shirt and a

Figure 6.10. Polar-mounted telescope for H-alpha observing. Note the movable screen for shading the observer. Robert Hess.

wide brimmed hat are advisable. Some observers choose to wear a long-billed baseball cap backwards, thereby covering the neck area when placing the eye to the telescope.

Polar-Mounted Heliostat

The heliostat is a device that normally uses two front-surfaced mirrors to reflect light to a stationary telescope mounted in either a vertical or horizontal position. Using a heliostat-fed telescope is a lot like studying in the science lab and looking through a microscope. This is a very comfortable means of observing.

California amateur astronomer Robert Hess has constructed over several years a lightweight, portable heliostat to feed his Televue TV-85 (600 mm focal length f/7) refractor and a tilting Daystar filter (0.5 Å bandwidth). The Hess' heliostat is not of the conventional two-mirror layout; rather, its design makes use of a single mirror in a polar mounted configuration. A single polar-mounted mirror requires only one movement to track a celestial body across the heavens.

The heart of the Hess mounting is a 4.25 inch minor axis 1/10 wave flat mirror that originally served as the diagonal in a large Newtonian reflector. The mirror mount swivels on bearings that are set within two rails that rigidly hold the mirror and telescope in polar alignment. Slow-motion controls permit adjustments in right ascension and declination. The Televue telescope is attached to the rails by its threads on the objective cell and a bolt underneath the clamshell holder. Two lightweight photographic tripods provide a very rigid and stable support for the system and the angle necessary for accurate polar alignment. Hess accomplishes the alignment during the daylight hours through the use of a compass and a digital level in four successive steps:

1. Use the compass to position the front (north) tripod.
2. Point the center post of this tripod toward the zenith using the level.
3. Position the rear (south) tripod so the rails point to true north using the compass.
4. Use the level to set the height of the rear tripod to within 0.1° of polar elevation.

While seated at the eyepiece of the telescope, an observer has easy access to the focus, filter tilt, tracking speed, R.A., and Dec. control knobs and altitude/azimuth fine-tuning devices that modify the mount's polar alignment.

Hess reports that he has been very pleased with the outcome of his custom project. His initial goal was to develop a portable, stable, and comfortable observing station for Sun watching. The fact that the telescope is lightweight, stores in two bags, and is easy to carry in one trip from the trunk of the car to a favorite viewing spot proves the convenience this setup offers.

Tilter Mechanism

Robert Hess's heliostat makes use of an end-loaded tilting H-alpha filter for rapid CWL changes. A front-loaded filter often has a tilting device incorporated in the mounting plate of the etalon. Anytime a light ray strikes a filter other than straight on, it creates a longer path through the filter. A longer path means a greater separation of the reflecting plates of the etalon, and for that ray, a change in the transmission characteristics of the filter to a lower wavelength, and a widening of the bandpass.

Solar observers desiring to view a Doppler-shifted event with a sub-angstrom filter take advantage of this characteristic by building various mechanisms that control the amount of tilt a filter is given. The basic principle relies on creating pivot points on the filter housing (or oven) that are 180° apart and instituting a spring loaded push-pull system to tilt the housing a controlled amount. The tilting mechanism usually has ends that accept a threaded adapter or other support equipment. Constructed from flat metal plates or cylindrical tubing, a tilter can be designed as a compact and attractive unit.

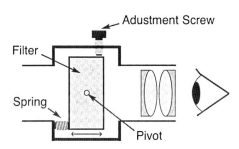

Figure 6.11. Schematic of a tilter for a sub-angstrom H-alpha filter.

The required amount of shift of the CWL is rarely more than a few angstroms, which translates to only several degrees from the perpendicular with most sub-angstrom filters. In the diagram, the rudimentary construction of a tilting device is illustrated. The pivot point will be positioned in line with the etalon, so that the filter rocks back and forth on its edge. The spring and push/pull screw could be located several possible ways, depending on the requirements of the designer. So long as the spring and screw work against one another, a controlled degree of tilt will be given to the filter.

References

1. *Solar Observing Techniques*, C. Kitchin, Springer-Verlag, 2002
2. *How to Observe the Sun Safely*, L. Macdonald, Springer-Verlag, 2003

Chapter 7

Monochromatic Solar Features

Prominences

Without a doubt the chromosphere is home to some of the most spectacular events that the amateur may study. In Chapter 1, we stated that as interior gas pressure decreases with distance from the Sun's core the magnetic field generated via differential rotation gradually becomes the dominant force. In the chromosphere, magnetic activity determines a flurry of seemingly out of control events. Many of these episodes release a fantastic amount of energy, sometimes ejecting material from the Sun at a truly astronomical velocity. As a result many amateurs find the chromosphere an extremely exciting place to examine. All the action is visible in the H-alpha light, although a few features can be better seen in the Calcium-K line.

One of the most widely studied chromospheric features is the prominence. A prominence (prom) is loosely defined as a cloud of gas suspended above the surface of the Sun. Typically a prominence has a temperature of 10,000 K and a density many times the surrounding chromosphere. Often a prom will outline the shape of its supporting magnetic field, changing appearance as the field evolves. Prominences, therefore, provide one means for professional astronomers to map a magnetic field on the Sun. The variety in size, shape, brightness, and motion found in prominences make them the most fascinating of phenomenon to watch and record.

Viewed at the Sun's limb a prominence is bright, but when seen against the disc it comes across as being dark and is then called a filament. The prominence and filament are one in the same feature, yet it took solar astronomers a time to recognize that. The brightness of a prominence at the limb is the result of viewing an emission feature before the cool, dark background of space. View a prom against the hotter, denser solar disc, and the filament appears as a dark absorption feature. A filament before the Sun's disc, if sufficiently energetic, can be a bright event, but as a rule it appears as a light gray to black thread. Filaments associated with a sunspot group in general are narrow, dark, and winding lines; the long, faint, thick variety of filaments are usually unaccompanied, changing appearance over time ever so slowly.

The one predictable factor regarding filaments and prominences is that anything is possible. A transient feature with a lifetime measured from minutes to several months, a prom can occupy space greater than the solar diameter, though it

J.L. Jenkins, *The Sun and How to Observe It*, DOI 10.1007/978-0-387-09498-4_7,
© Springer Science+Business Media, LLC 2009

is often much shorter in length, closer to several thousand kilometers. Not limited to the sunspot zones, a filament can appear in the higher solar latitudes and then, because of differential rotation, several structures may align in an east-west direction, seeming to link together and forming one long strand of filamentary material. This configuration when formed is called a polar crown. Viewed at a steep angle from Earth, a filament often presents projections on one side that look like a "scalloped" edge. These projections from a filament extend to the area where an active region's magnetic field reverses polarity, an area called the neutral line.

There is a significant connection between proms and filaments, active regions, and solar flares. A filament develops where conditions result in the elevation of dense gas above the surface of the Sun. One of these conditions is seen as a solar flare manipulates material through thermal effects; the result is the condensation of gas in the lower corona, and this gas then rains down onto the surface. Material can also be ejected from the Sun through the most violent energy releases, the solar flare. Lastly, and related to all proms and filaments is the shouldering effect of gas by a magnetic field poking into the chromosphere from below.

Two broad groups are used to classify prominences and filaments: quiescent (quiet) and eruptive (active). For the most part a quiescent prominence behaves in a calm manner, changing its appearance only moderately with time. An eruptive prominence, because of its bursting nature, is, of course, the more exciting to observe. Since the distance to the Sun is great, and although some prominences are quite large (up to hundreds of thousands of kilometers in length), real-time activity does normally appear gradual. However, don't think that an eruptive prominence necessarily moves at a snail's pace; dramatic changes can easily be seen within a short span of 60 s. Quiescent examples do not always remain quiet and unassuming, either; they can become disturbed and erupt toward space, disappearing completely from the previously occupied region. And because a prom/filament has disappeared, don't conclude the show is over; a feature may reappear minutes or days later in the same area.

The hedgerow is a typical quiescent prominence. It gets its name from the obvious similarity in appearance to rows of trees and hedges left as windbreaks near farmers' fields. The ends of a hedgerow prominence are often magnetically anchored to the surface below. Normally a stable, quiet feature, when one end of the hedgerow breaks free from below, this is taken as an indicator that the prominence is about to erupt. Other frequent shapes that a quiescent prominence may assume include the suspended cloud, the mound, the loop, the tornado, and the tree trunk.

Prominence loops are particularly beautiful limb structures. Following the outburst of an intense flare, the solar material ejected into the corona begins to cool and condense, flowing back onto the underlying surface. Sometimes the material flow is down both sides of a loop, and occasionally it can be seen rising up one side and returning down the other. Should the loop be open and incomplete, a feature called coronal rain can be observed, consisting of faint streams and knots of gas pouring back through the chromosphere. These effects are clearly evident over time, especially in a video clip that accelerates the material flow.

When a prominence erupts, it does so away from the Sun. Eruptive-type prominences have descriptive names that are indicative of their appearance. The surge is a controlled or straight-line eruption of material following a powerful flare. A surge will often burst outward, lose inertia, and fall back onto itself with a great splashing effect. Should the surge be in disarray and uncontrolled, with material going in every direction it's called a spray. Another form taken is that of a long, narrow column of material known as a jet or an impulsive prominence. A typical surge velocity is approximately 150 km/s; a spray regularly exceeds 200 km/s. The amateur astronomer observing these limb features can make a line-of-sight velocity measurement by noting the changing position of a knot of gas in the prominence against a time scale.

A filamentary surge viewed on the disc of the Sun is, by and large, moving in the direction of Earth and will appear Doppler shifted toward the blue wing of the H-alpha line. The Doppler effect results from the compressing or stretching of light waves because an object is rapidly moving toward or away from the observer. Sound waves represent a perfect example of Doppler shifting by their change in pitch as a train whistle nears and passes a station. To the observer of a Doppler-shifted eruption, the filament will begin to fade, gradually becoming invisible until it has disappeared completely. By tuning the CWL of an H-alpha filter toward the blue portion of the spectrum the prominence will reappear, unless it has been completely obliterated. This event, the sudden disappearance of a filament, is called a disparition brusque. Another interesting but uncommon effect involves an impulsive-type prominence near the limb that has a trajectory toward Earth. In this case only a portion of the prominence remains visible when the H-alpha filter is on-band. As the filter is tuned toward the blue wing, the lower region disappears while the upper parts become visible. Apparently the outer regions are experiencing a greater velocity than the lower region.

There is no system of classification that can describe all the forms a prominence can assume. Regardless a number of schemes have been devised over the years; one interesting attempt was by D. Menzel and J. Evans in the early 1950s. This methodology (Table 7.1) is similar to the McIntosh sunspot classification scheme in its use of a three-letter designation. In the Menzel-Evans classification, the first letter represents the place of origin for the prominence, whether it is descending from the corona or ascending from the chromosphere. The second letter tells us if the prominence is related to a sunspot or not, and the third letter is a description of the prominence's appearance. The delicate loops pictured in Figure 7.1 are from April

Table 7.1. The Menzel/Evans classification system for prominences

A – Prominences originating in the corona (Descending)

S – Spot prominences	N – Nonspot prominences
a. Coronal rain	a. Coronal rain
f. Funnel	b. Tree trunk
l. Loop	c. Tree
	d. Hedgerow
	f. Suspended cloud
	m. Mound

B – Prominences originating in the chromosphere (Ascending)

S – Spot prominences	N – Non-spot prominences
s. Surge	s. Spicule
p. Puff	

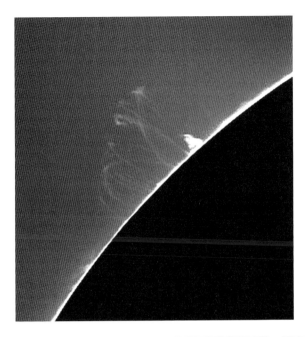

Figure 7.1. Three forms of prominences are visible in this photo. Faint delicate loops, coronal rain, and a surge were active in 2001. Jamey Jenkins.

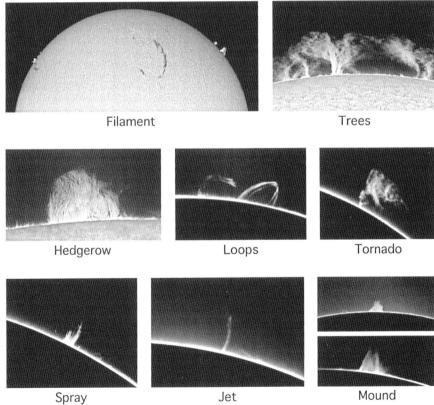

Filament

Trees

Hedgerow

Loops

Tornado

Spray

Jet

Mound

Figure 7.2. A sampling of prominences. The filament on the disc, the trees, and the hedgerow are courtesy of Eric Roel. The rest from Jamey Jenkins.

Figure 7.3. High-resolution image of a filament from June 2002. The underside presents a scalloped edge where the prominence extends to "anchor points" in the lower chromosphere. Gordon Garcia.

2001 and have a Menzel-Evans classification of Asl. These prominences are originating from above so as to be descending; hence, the "A" classification. There is a sunspot group just appearing around the limb (not visible in the photo) to which these prominences are associated; therefore, there is an "S" for the second letter. Being in the form of loops, the third letter in the classification is an "l." The bright surge just to the north of the loops would be classified BSs. The coronal rain located between the two, and originating farther into the corona, would be ASa.

Solar Flares

Novice observers and the general public often incorrectly believe that prominences and flares are the same phenomena of the Sun. A cloud of gas suspended above the surface of the Sun aptly describes a prominence; a solar flare is the swift release of energy that has accumulated within the magnetic field of an active region. A flare will show itself as an unexpected brightening of about twice the intensity of the surrounding chromosphere. Changes in brightness because of flaring can also be seen in the area around an active region known as a plage (pronounced PLA-juh). Visible in the lower chromosphere and routinely seen well in the light of Ca-K (393.3 nm), a plage can surround a sunspot as a cloud-like form with no particular consistency in brightness or shape.

A plage marks where the magnetic field associated with a spot is located. It is the chromospheric version of the photosphere's facula, and when discovered in the late nineteenth century were called bright flocculi. A generalized term, flocculi was used to describe many features of the Sun's chromosphere at that time. For example, filaments were known as dark flocculi. The area in a plage lacking concentration is called the plage corridor and denotes the location of a reversal of magnetic polarity. It is important not to confuse a plage with a solar flare. A plage is not as intense as a flare and is stable for a longer period of time. A flare is by far the more transient of the two events.

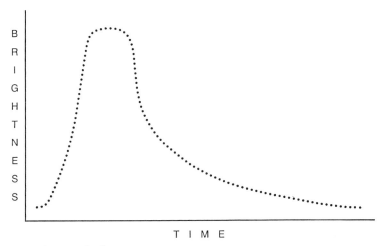

Figure 7.4. Light curve of a flare.

Nearly all solar flares will be observed in the light of H-alpha, because it is here that a flare is in emission in the visible wavelengths. The first seen flare was intense enough to be visible in white light (see Chapter 4), but with the invention of the spectroheliograph the frequency and magnitude of flare activity became evident. A well-developed sunspot group of McIntosh classification D, E, or F is particularly notorious for producing solar flares. Motion experienced by a sunspot or the emergence of new magnetic flux in a sunspot initiates many flares. A twisting and distortion of the neutral line within a complex sunspot group often results in flare output. The solar cycle is also indicative of their frequency, with very few flares occurring near minimum. At maximum the daily occurrence of an event would not be unusual, and larger events typically occur about 2 years after sunspot maximum. A flare can and does exhibit dramatic changes minute to minute; usually the event ends within an hour.

Figure 7.4 illustrates how the light curve of a typical solar flare might appear. The flare usually begins with a rapid brightening of several points, or kernels, located inside or near a sunspot. These grow in size and brightness; for a large flare it may take but several minutes, with a small flare taking an even shorter time span. The period of rapid brightening is called the flash phase, which is then followed by a gradual, slow retreat in brightness. If the flare is an energetic one, its appearance may develop into a double ribbon-like structure that marks the location of the neutral line (Figure 7.5).

Local solar temperatures in the flare's vicinity spike to many millions of degrees and may erupt a filament that has developed above the neutral line, resulting in a disparition brusque. The formation of one or more loop prominences is a certainty because of the flare's intense heating of the lower corona. A flare is a dramatic event!

Several other interesting phenomena occur because of flare activity. One feature that can be rather challenging to spot is the Moreton wave. Particularly large explosive flares release shock waves that expand out across the disc of the Sun, assuming the shape of an arc. The expanding velocity of a Moreton wave is approximately 1000 km/s, turning it into a rapidly moving wave front. Viewed on-band in H-alpha expect to see a Moreton wave as a lighter, low contrast feature; off-band in the wings of H-alpha,

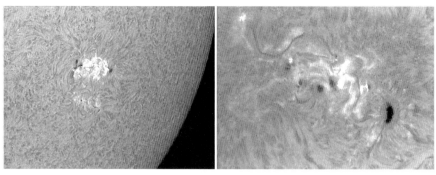

Figure 7.5. Solar flares in action. These photos from May 2, 2004 (*left*), and July 5, 2003 (*right*), were obtained with a Daystar 0.38-angstrom H-alpha filter. Gordon Garcia.

it appears darker. As the wave crosses the solar disc it may encounter a filament that will react like a person in the ocean encountering an incoming wave – to first be lifted upward and then deposited back on the ocean floor. The filament will appear to blink in and out of visibility, as it is Doppler shifted in and out of the bandpass, a consequence of the rapid up and down motion caused by the passing wave.

The coronal mass ejection, abbreviated as CME, is sometimes the result of solar flare activity. A huge bubble of coronal gas flows out of the Sun and becomes a shock wave in the solar wind. Developing over a period of several hours a CME occasionally becomes larger than the Sun's disc. The SOHO satellite has, for the most part, continuously patrolled the corona since 1996. Time-lapse movies assembled from SOHO images dramatically show the expanding nature of CMEs. SOHO movies are available at several Internet sites for viewing.[1]

The optical classification of a solar flare is derived from the heliographic area of the flare at maximum brightness in H-alpha. Also known as the "Importance Classification," it is a two-digit system, the first number representing the estimated area of a flare in square degrees and the second digit a measure of the flare's brightness. Area is assigned a rating from 0 (the smallest of flares) through 4 (the largest). Some variations on the scheme use an "S" for the tiniest of regions, designating them as a "sub-flares."

The brightness of a flare is assigned F, N, or B (faint, normal, or bright). As an example, a solar flare may be classified in this system as: Importance 2N, Importance 4B, etc. Flare classification using this system is a bit subjective, particularly as a region nears the solar limb, where foreshortening of the feature makes estimation difficult. Even when a flare is near the solar meridian a "gray area" exists among the terms faint, normal, and bright and their interpretation by an observer. Amateur astronomers desiring to classify flares from their observations will use the Importance Classification system because it is based on criteria they can measure (see Table 7.2).

A solar flare may also be classified according to the peak flux it causes in X-rays. Measured by the Geostationary Operational Environmental Satellites (GOES), this classification follows a scheme of A, B, C, M, or X, each sub-divided further into ten numerical categories (0–9). The GOES system stems from the relative X-ray brightness of a flare in comparison to the X-ray brightness of the remainder of the Sun. As an example of this system in action, an M4 Class flare is seen as ten times as powerful as a C4 Class flare, and an X4 Class flare is ten times more powerful than the M4 Class.

Table 7.2. Optical solar flare classification

Importance 0 = < or = 2.0 heliographic square
  degrees
Importance 1 = 2.1–5.1 square degrees
Importance 2 = 5.2–12.4 square degrees
Importance 3 = 12.5–24.7 square degrees
Importance 4 = > or = 24.8 square degrees
Sub-code for brightness – F (faint), N (normal), B (bright)

The NOAA has recently formulated a scale that measures the amount of energetic particle flow from a flare reaching the Earth. The scale reads as S1, S2, S3, S4, and S5, with S1 being a minor flare event and S5 an extreme flare. Again, each step in the scale represents a tenfold increase in particle flow. It is hoped that this scale will be more "user friendly" for the general public when estimating space weather conditions and public safety.

Chromospheric Network

Covering practically the entire Sun, the chromospheric network is a structure resembling fine latticework. Viewed in the light of Calcium at 393.3 nm (Ca-K), the network is a bright web-like pattern. In H-alpha the tones become reversed, with the center of the pattern appearing bright and the webbing dark. This network is composed of coarse and fine mottles, the coarse mottles being up to 20,000 km across and sometimes merging to form plages. The fine mottles are a few hundred kilometers wide and can be several thousand kilometers long.[2] Magnetic alignment with the supergranules in the underlying photosphere causes the chromospheric network. Unlike the granulation of the photosphere, the network's cells have a diameter of nearly 40 arc seconds, each with a lifetime of about a day.

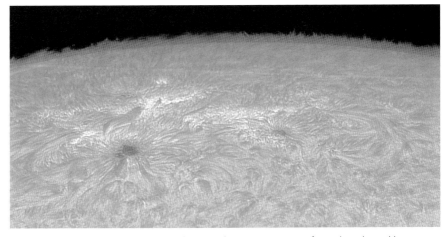

Figure 7.6. This image from Eric Roel illustrates the tenuous nature of spicules. The jet-like appearance at the limb is complemented by the whirled effect created by the magnetic field of the sunspots.

The pink-colored ring surrounding the edge of the Sun at a total eclipse is the approximately 2000-km-thick chromosphere. Earlier telescopic observers of eclipses thought this edge ring resembled a multitude of "gas-jets" shooting out from the Sun. Given sufficient resolution, the chromosphere reveals itself to be composed of many thousands of these fine structures, called spicules. Seen at the limb as a bright emission feature, a spicule viewed against the disc appears dark and is known as a fibril or mottle. Collectively the spicules outline the chromospheric network, giving its edge a fuzzy, somewhat hairy appearance. Rising to an average height of 7500 km and having a diameter of about 800 km, spicules extend from the limb at an angle of 70–90°. The lifetime of a spicule is approximately 5 min, ending its existence by falling back into the Sun or disappearing from view.

Groups of spicules assume various shapes and patterns. If they are observed close to the limb but not on it, the hairy appearance is more evident. Clusters of spicules are sometimes referred to as a brush. Under good seeing conditions the fine spicules of a brush give a striking three-dimensional impression. When the spicules are aligned to form a column or row, they are known as a chain. One of the aesthetically pleasing shapes is that of a rosette, the radiating of spicules from a central point, like a flower petal. An active region containing sunspots with a strong magnetic field has influence over the nearby spicules. Any located adjacent to the sunspot group become bent and stretched, following the local magnetic field lines, giving the illusion of a whirlpool.

When the thin gas in the upper chromosphere adheres to a local magnetic field, finger-like projections can be seen that indicate the location of lines of magnetism. Remember the science class experiment with iron filings and a bar magnet? The same magnetic alignment principle is in effect on the Sun. These projections are called fibrils. Many times you will find fibrils connected to a larger filament/prominence, along the feature's scalloped edge. Some fibrils exhibit dimensions exceeding a length and width of 10,000 km and 2000 km, respectively.

An emerging flux region is an area of new magnetic flux making its way into the photosphere, such as at the edge of a sunspot. Often tiny bright points or miniature flares less than 5 arc seconds across appear there for several minutes to a few hours. These bright points are called Ellerman bombs and are clearly observed in the wings of H-alpha, because they are located in the lower portion of the chromosphere. Also known as moustaches, they may be related to the reconnection of magnetic fields. The cause is not clear.

Monochromatic and white light observing of the Sun complement one another because the magnetic fields that create activity in the photosphere and the chromosphere are directly related. If you wish to gain a fuller understanding of what is happening within an active region on the Sun, you should follow the various stages as they occur, in many cases from the solar surface outward. To do this requires monitoring the white light and monochromatic Sun.

Observing Projects

The white light section of this book highlights several possibilities for an amateur astronomer desiring to become involved in a worthwhile observing program. The several benefits of "observing with a purpose" include: improving your skills; developing new friendships; and the primary goal, making a contribution to science. The monochromatic observer likewise is encouraged to develop an observing program.

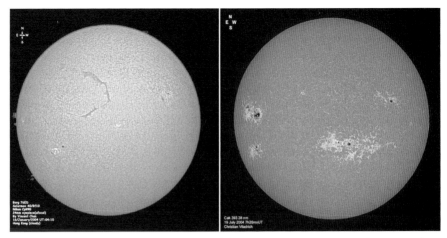

Figure 7.7. Whole disc patrol photographs. The *left image* is in H-alpha light and is by Vincent Chan. The image on the *right* is by Christian Viladrich, in light from the Calcium K-line.

Again photography is preferred over sketching as the recording medium best for morphology work. Whole disc and active region photography are areas full of opportunity for a properly outfitted solar observer. Accurate placement of features on a drawing blank can be difficult at best, and it's a nearly impossible task to draw all that can be seen at the eyepiece. Not a totally wasteful procedure, sketching has its place for note taking, making brightness estimates, and so on. Photography, whether conventional or digital, is the only reliable means of capturing a one-time only event. But image capture isn't the only avenue for exploration of the monochromatic Sun; computing the line-of-sight velocity of proms at the solar limb is one of a number of other options. Cataloging and classifying solar flares consumes some observers (Figure 7.7).

Solar Morphology

If you are interested in pursuing this area of study you can begin by re-reading Chapter 5 and the section dealing with morphology programs for the white light observer. The advice with respect to image orientation, the recording of data, and the usage of your photos can be applied to chromospheric studies. The significant difference between white light and narrow band work will be the bounty of additional detail in the monochromatic photo. Time requirements may vary, depending on the project being worked. For example, photographing the life of a white light sunspot can take a number of days, whereas energetic chromospheric activity may last 10 min to several hours. With this time frame to work with it is a near certainty that the observer can record an entire event without interruption.

Whole disc photos are useful when illustrating the condition of the Earth-facing chromosphere at the time of day in question. The goal of any whole disc photography program is to obtain sharp, clear, wide-angle pictures of the Sun. The photograph should have a standard format, so that a legitimate comparison can be made between it and other white light or monochromatic patrol pictures.

A whole disc photo on average exhibits a resolution of about 2–3 arc seconds. However, because the monochromatic observer is photographing through a narrow band filter, atmospheric dispersion and poor seeing will be minimized, to the photographer's advantage. Some instruments create a "sweet-spot" of transmission; a whole disc photo is best obtained with an assembly that fits the entire 32 min of arc solar diameter inside this area. A short to medium focal length telescope using a front-loading filter is an excellent tool for this work. An end-loading filter works well when coupled with a suitable telecentric or when stopped on-axis to the nominal f/30 focal ratio. Aperture and consequently high-resolution are not the ultimate concern with this type of photography, since the purpose of the project is to depict only the position of features on the disc. What a photographer will seek to accomplish is even illumination across the solar disc and a minimum of the "sweet-spot" effect.

Active region photography is a bit more challenging because the goal is to obtain a close-up image of the Sun with a resolution of 1 arc second or better. High-resolution work magnifies atmospheric seeing conditions and instrumental defects such as those from vibrations. Practice is key to developing good technique, and perseverance by the observer, at even the poorer sites, will uncover some moments of excellent seeing. A relatively short duration event involving a solar flare, erupting prominence, Moreton wave, or other transient happening can make for an exciting series of photos. A movie clip can be assembled from these images that spectacularly demonstrate the event's life in a compressed time span (Figure 7.8).

Prominence Measurements

Analyzing a photograph or carrying out a visual observation that leads to the measurement of a feature is a simple task. Why would an amateur be interested in determining the position of a prominence or the height of a hedgerow formation? Determining the position of a prominence is helpful for identification purposes. Active regions in the photosphere are typically long-lived features that are cataloged in a timely manner. On many occasions a chromospheric feature is related to an active region, but at other times it appears as a randomly placed event that comes and goes within a few hours.

Line-of-sight measurements mean the apparent rather than the true position, size, or velocity of a feature as seen from our station on Earth. The prominence measurements discussed here are based on line-of-sight observations. Prominences, when compared to filaments, show their true profile because many limb events happen at an angle of approximately $90°$ relative to the observer. The estimated height of a filament is obviously far easier to determine on the solar limb than at the center of the Sun's disc, where an observer is positioned directly above it. Expect the measurement to be compromised to some degree by instrument quirks or limitations imposed by the atmosphere. However, for the amateur estimating the height or eruption velocity of a prominence, satisfying a curiosity about the feature is what's important. It's educational and interesting to know that a feature under study is many times the size of Earth and perhaps is being ejected from the Sun at a truly astronomical speed.

The use of a Stonyhurst Disc for finding the heliographic coordinates of a disc feature is feasible if a whole disc photo has been secured in which an accurate east-west line is established, as outlined in Chapter 4. Analyzing and graphing

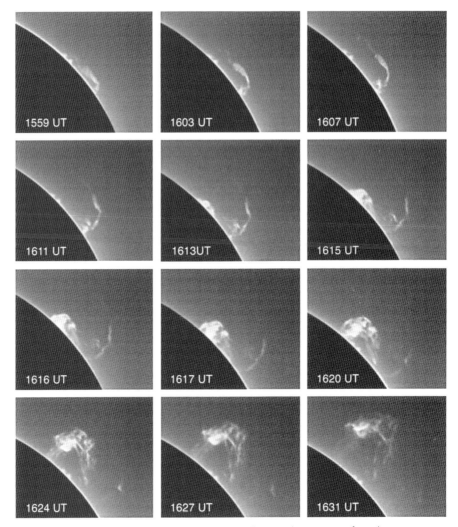

Figure 7.8. Series of photos from March 5, 2000, showing the eruption of a solar prominence. Universal Time appears in the *lower left hand* corner. Steve Rismiller.

coordinates to form a butterfly diagram is an instructive project for a monochromatic observer interested in activity center movements throughout a solar cycle, as well as detecting the differential rotation of the Sun.

The position angle (P.A.) of a prominence is a convenient and commonly used means of identification. On the Sun, P.A. is measured from the north moving eastward (counterclockwise) and divided into 360° around the limb. The cardinal directions, therefore, are as follows: N=0°, E=90°, S=180°, W=270°. A clear cell template can be created with a diameter equal to the standard disc size (15–18 cm) serving as an overlay for a monochromatic photo and enabling the P.A. to be read directly. If the template is printed black on a clear background, a negative print of the Sun makes reading the divisions easier. As discussed in Chapter 4, the photo must have an accurate indicator of celestial east-west, normally found by using the drift method. The template is positioned on the photo in accordance with the

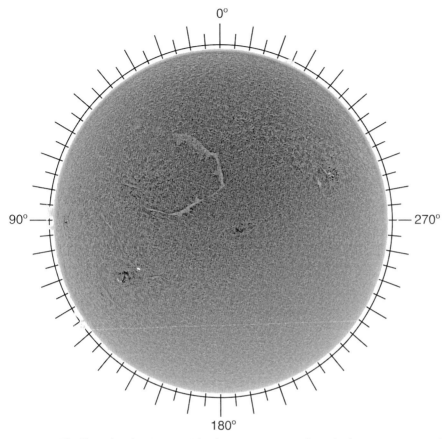

Figure 7.9. The Chan photo from Figure 7.7 has been negative printed to make the prominences stand out against the protractor overlay in this illustration. Each tick mark represents 5° of position angle around the limb of the Sun. Prominences are visible at 87°, 93°, 126°, 270° (faintly), and 336°.

celestial directions and then tilted by the amount of P as indicated in the daily almanac. The P.A. for a feature on the Sun's limb is then read directly from the overlay using this method.

An even more direct approach, although perhaps less accurate, is to use a protractor reticule with an eyepiece that lets the whole disc of the Sun be visible. A reticule of this type is built into a few astrophotography-guiding eyepieces and is available through optical houses or from surplus dealers. At the telescope, a reticule must first be aligned to celestial east-west by again using the drift method. Once a thread (remember which thread, because they all radiate from the center) is aligned east-west, resume the driving motion of the mounting and slew the telescope so the Sun is centered in the protractor.

At this point you might want to have a blank drawing form prepared that has a lightly printed representation of the reticule on it. Rather than drawing pictures of the prominences, simply sketch squares or rectangles on the edge of the circle representing the position and approximate size of the prominences as seen with the protractor scale in the eyepiece. After all the prominences visible have been sketched, start at the north pole and go around the Sun toward the east. Label the proms numerically beginning with 1; this serves to identify the prominences that are photographed. The P. A. is read directly from the sketch, and correction is made for the daily P value. If you are careful, a position can be identified within a couple degrees. This method is most suitable for a photographer doing high-resolution work acquiring the P.A. of a prominence, or if you are just interested in studying the frequency of prominences at various latitudes.

Determining the height of a prominence above the solar limb with a whole disc photo is a procedure based on the known diameter of the solar disc. From Table 1.1 notice that the Sun has a diameter of 1,391,980 km (its radius is 695,990 km). For comparison Earth has a diameter of 12,756 km.[1] With a whole disc photo of the Sun (the larger the photo, the greater the accuracy), use a finely divided millimeter ruler to measure both the diameter of the Sun and the height of the prominence. Divide the prominence measurement by the disc measurement and then multiply the resulting quotient by 1,391,980 to find the height of the prominence in kilometers. Of course, this method's accuracy will depend on the resolution of the photograph; a higher quality measurement requires a higher resolution image.

A visual method of determining prominence height requires the amateur to possess either a graduated reticule or a filar micrometer. A timepiece is used to calibrate the tick marks of the reticule or the vernier scale of the micrometer. Because of the varying distance to the Sun throughout the year the calibration procedure must be performed whenever a measurement is taken. Begin by orientating the measuring scale in an east-west direction by observing a solar feature's path across the eyepiece field with a magnification of 100× or greater and the drive mechanism off. Once the scale is parallel to the drift, position the Sun in the telescope so that the west limb is tangent to a thread perpendicular to the scale, turn off the drive, and time the transit of the entire solar disc across this thread. Repeat the procedure several more times and average the results.

Now return the Sun to the tangent position as above, and time the interval of the west limb's edge drifting between two threads of the scale. Repeat this several more times, and average the results. Divide the whole disc transit time by the drift time between two threads to obtain the disc-to-scale factor. Divide the known diameter of the Sun (1,391,980 km) by this factor to find the image scale, which is now the number of kilometers each division of the scale represents. So long as the diameter of the virtual image of the Sun in the telescope doesn't change with respect to the graduated scale, the image scale will remain the same. For improved accuracy, use a short focal length eyepiece, which provides greater magnification. Once the reticule or vernier is calibrated, it's possible to position the scale over a prominence, count the number of divisions consumed, and multiply that by the image scale to find the line of sight height of the prominence above the limb (Figure 7.10).

Determining the line-of-sight velocity of an eruptive prominence requires obtaining a series of photographs that are accurately timed and show a minimum of two points of reference located on or near the solar limb. Reference points could include two quiescent prominences or a quiescent prominence and a sunspot. Knots of gas within the erupting prom are carefully identified in each photo; these are then measured for their displacement relative to their previous position while noting the amount of time between successive photos. The scale of the photos (km/mm) must be known or the center of the Sun visible in each photo to determine the scale.

The basic technique can be demonstrated by the following procedure. Having obtained a photo series that meets the above criteria, arrange them on a table

Figure 7.10. Filar micrometer used for prominence measurements. Jamey Jenkins.

before you. Use tracing paper over the first photo and mark with a fine point pen or pencil the reference points, the solar limb, and the center of the gas knots you will be measuring (if the scale is not known, also mark the position of the center of the solar disc). Transfer the tracing paper to the second photo of the series and carefully align the trace to the reference points; again mark the position of the knots of gas of the eruptive prominence. Repeat this procedure until the last photograph is measured. Identify the knots on the tracing paper with numerals (1, 2, 3, etc.) that correspond with the outward progression of the prominence.

Since the scale in kilometers of the photo is known, use the millimeter ruler to measure the distance between any two successive knots to find the distance the prominence has been ejected. Line-of-sight velocity (V) is an expression of $V = D/T$, where D is the distance between two points and T is the time interval between those points. Therefore, the expression of velocity would be so many kilometers per second. Outward velocity, away from the Sun, has a positive (+) value; incoming material, falling toward the Sun, has a negative (−) value. Without knowledge of the print scale of the photographs, but with a field of view capturing the center of the disc, the radius of the Sun may be given consideration to obtain the scale. The solar radius from Table 1.1 is 695,990 km; therefore, divide 695,990 by the distance from the center of the photo to the limb as obtained with the millimeter ruler to find the kilometer/millimeter scale of the photos.

References

1. *How to Observe the Sun Safely*, L. Macdonald, Springer-Verlag, 2003
2. *Solar Observing Techniques*, C. Kitchin, Springer-Verlag, 2002

Chapter 8

Solar Photography

A Hobby Within a Hobby

Most solar observers record an observation photographically for one of two specific reasons: to either archive what has been seen and perform measurements that are difficult to make at the eyepiece, or to share what they have experienced with friends and the world. When aptly applied, photography is an ideal means of replicating what can be seen at the eyepiece; details intermittently glimpsed are frozen, enhanced, and the clarity of an observation improved.

In the opening paragraphs of this book, we spoke of astronomical photography as being a "hobby within a hobby." At one time the complexity of photography necessitated the development of multiple skill sets for an amateur astronomer. Dubbed an astrophotographer, the amateur pursuing photography had to master darkroom work, acquire an uncanny sixth sense regarding atmospheric seeing conditions, and possess an understanding of optical principles. Months could be spent acquiring sufficient knowledge of your astronomical interest, and then developing the knack for successful photography of it.

Times have changed! For the majority of us, the basement or closet darkroom has been returned to storage space for the family. Processing photographs these days doesn't necessarily mean splashing around in chemical trays under the illumination of a red safelight. Instead, all the tweaking of a photo is done from your desk or laptop computer. A click of the mouse here, a keystroke there, and like magic, photos are printing from a compact unit on the table beside you, or are prepped for posting on the web for all to see.

However, astrophotography is still a hobby within a hobby, and although different skill sets need to be developed, these are crafts that the average techno-junkie has often already scrutinized. Today's science-minded individual understands the home computer, web cam operation, the email network, and file manipulation tasks. Still, it does help to have a basic knowledge of optics, but with today's ready-to-go telescope assemblies, even that is becoming less of a necessity.

A Brief Historical Perspective

The light-sensitive properties of certain chemicals were known long before the Frenchman, Niépce, exposed sensitized paper to light in creating the first successful photograph. Niépce's experiment required a tedious exposure of 8 h to record the legendary window scene. Not long after Niépce, around 1839, another

J.L. Jenkins, *The Sun and How to Observe It*, DOI 10.1007/978-0-387-09498-4_8,
© Springer Science+Business Media, LLC 2009

Frenchmen, Louis Daguerre, perfected the Daguerreotype process. A copper plate coated with silver iodide, exposed in the camera and developed with mercury vapors, is called a Daguerreotype. High-quality photos for that era were now possible, but the Daguerreotype was a very expensive process, and only the original photo was produced. There was no intermediate negative available with the Daguerreotype for creating additional copies.

Around this time John Herschel, astronomer and son of the famous William Herschel, coined the term photography, to describe the new invention. Photography literally means "light writing." In the modern digital photography vernacular, a photo is often referred to as an image; photograph and image have become interchangeable terms. Also around this time, William Henry Fox Talbot created another photographic process dubbed the Calotype. The paper negative produced by the Calotype could be used to generate positive prints. Definitely the ability to reproduce copies of a photo was a distinct advantage of the Calotype over the Daguerreotype process.

Solar photography had its beginnings in 1845, when Léon Foucault obtained the first photo of the Sun. This first picture was detailed enough to show limb darkening and a handful of sunspots. Foucault is primarily associated with the test bearing his name, used for judging the quality of an optical surface.

The sensitivity of an emulsion (light-reactive coating) in these early achievements was weak and required laboriously long exposures, that is, until Fredrick Archer devised the wet collodion process. Exposure times with Archer's plates were now reduced to only a few seconds. The wet process, while faster, had the disadvantage of requiring a coating and processing lab at the photography site. Wet plates were to remain wet until the fixing of a photograph was complete. Also, about this time (1870), Charles A. Young succeeded in photographing the first prominence outside of an eclipse.

A giant leap forward was realized the following year when Richard Maddox introduced a dry photographic plate. Maddox found that silver bromide could be suspended in a layer of gelatin, this substance then used to coat a photographic plate. The development of the dry plate led to the convenience of storing and processing plates at the photographer's discretion. Shortly before this time, the physicist James Clerk-Maxwell had demonstrated that color photography was feasible. By taking separate exposures through red, green, and blue filters Clerk-Maxwell reassembled them into one photo using three slide projectors shining through the same three original filters.

George Eastman introduced the Kodak camera, and photography became available to the masses. Previously only trained individuals knew and understood the wonders of the photographic process. In 1889 Eastman introduced a flexible roll film instead of a paper negative in his cameras. Eastman Kodak continued to be a leader of ever increasingly sensitive emulsions throughout the last century. Techniques such as hypersensitization, which combat the inefficiency of an emulsion to weak light, were fully explored and implemented by photographers.

At Bell Labs in 1969 the charge-coupled device (CCD), invented by George Smith and Willard Boyle, quietly ushered in a new era in photography. Created with the intention of storing computer data, the new device was sensitive to light, and within 5 years the first imaging chip was constructed.[1] The following year Kodak assembled the first CCD-based still camera. Then in 1991 Kodak released a professional digital camera system for photojournalists, a modified Nikon F-3 that was equipped with a 1.3-megapixel sensor.

The initial camera marketed to the general public for home computer use was the Apple QuickTake 100 around 1994. The QuickTake featured a 640 × 480 pixel CCD and stored eight images in its internal memory. The development of the astronomical CCD camera followed rapidly, as well as the revolution in videography, the digital point-and-shoot camera, the digital single lens reflex (DSLR), and the computer web-cam.

The twenty-first century amateur solar observer uses these various digital devices and to a lesser degree film to record solar activity. The digi-revolution is continuing to evolve as newer cameras and support systems are introduced and current ones become outdated. By digital camera we mean any electronic device with a light sensor that is capable of producing a photographic image. When speaking of a detector, either photographic film or a digital sensor applies. In the following pages we will explain how current products and methods have been merged with techniques from the past to produce the high-quality images recently seen in amateur circles. The concepts and principles reviewed in these pages are applicable to whatever process is in use. The goal in each case is to capture sharp, detailed photos of the Sun with as much clarity as possible.

Solar Photography Basics

"What's the best camera setup for taking pictures of the Sun?" That's the question every veteran solar observer eventually hears from the newbie. When film was the only medium available to the amateur astronomer, this question was a fairly easy one to answer. At that time only a handful of camera models had features that excelled for astronomical photography. Desirable qualities included, among others, vibration free operation and a lightweight camera body.

Today, when responding to the same question, the particular needs for solar work remain the same, but when you put the digital camera in the mix, a far greater number of makes and models, some not even resembling a conventional camera, can be considered useful.

It is possible to hold any camera, digital or film, with its lens focus set to infinity up to an eyepiece and click away, acquiring on occasion a somewhat satisfactory picture. Images using this technique, however, are rarely successful, but it is possible. Rather, as an astrophotographer, you should seek a route that produces predictable results. Time and observational opportunities are rare and should not be left to chance. A photographic system fitting the following parameters will produce high-definition Sun photos.

To begin, a solar telescope must first be properly collimated. The poorly collimated telescope suffers from exaggerated aberrations and a loss of resolution. Even a refractor, which rarely needs adjusting, must be periodically checked for alignment. Sometimes only a tweak of a collimating screw can improve the performance of a telescope dramatically. Various wrenches and tools for collimation are available, Consult your telescope owner's manual or a telescope-making handbook for detailed instructions.

An experienced photographer will select an effective focal length for his or her system that provides an image with a scale sufficient to optimize the resolution of the telescope and detector combination. Often what you see in amateur circles is an image formed by a telescope and camera system that is too small for high-definition work, and resolution of a feature becomes compromised.

In the digital arena, the Nyquist theorem states that to be effective, a minimum of two pixels (picture elements) of an electronic sensor must cover the theoretical resolving power of a telescope. For telescopes in the 100–200 mm aperture range (theoretical resolution of 1.1–0.6 arc seconds) this translates into a sampling of about 0.5–0.25-arc seconds per pixel. Put another way, Nyquist is saying that to optimize resolution, the image placed on the sensor must be large enough so that at least two pixels are covered by an angular displacement in the sky equal to the theoretical resolution of the telescope. Over sampling the image a bit, meaning covering more than two pixels, is acceptable, and most experienced photographers recommend using a value near .1-arc second per pixel. Knowing the pixel size of your detector is vital; this information is usually published in the camera's owner manual or can be found on the Internet. The formula for working out an appropriate sampling for a telescope/sensor system is:

Pixel size of sensor (microns)/Focal length (mm) $\times$ 206 = arc seconds/pixel.[1]

A camera for astronomical use must have a means of accurately finding focus and monitoring image quality or, more specifically, the seeing conditions. If a digital camera is used, this likely will mean connecting to an external video monitor. Atmospheric cells of warm or cool air and accumulated warmth from the Sun resulting in gradually expanding components de-focus a telescope. Check and correct the focus of the telescope/camera system throughout an observing session. Setting up equipment should allow time for the telescope to acclimate itself to the surrounding air temperature before beginning an observing session. Depending on the difference in temperature, this could mean several minutes to an hour. A backyard observatory could be opened for a time prior to observing, so that air might circulate around the instruments.

The camera must not contribute vibration that could result in a blurred image. An electronic remote cord for a digital camera or an air-activated cable release removes the photographer's hand from the camera, eliminating an opportunity to introduce vibration. Some cameras are constructed with a low vibration shutter mechanism that helps in this area. A fast shutter speed is always preferable for high-definition photography, at least 1/125 of a second in order to freeze image motion caused by seeing conditions.

To facilitate the high shutter speed for white light photography, the intensity of the light reaching the image plane can be controlled by a photographer's selection of filters. A special thin photographic objective filter can be obtained that transmits a greater amount of light, requiring a shorter exposure time. Whereas the normal objective filter for visual use has a photographic density of 5.0, the special photo-version can have a density ranging from 2.5 to 4.0. Many reflex cameras permit an additional snap-on or screw-in filter between the eye and camera-viewing lens; this is where you should insert an additional piece of the photo-version filter material and so the system becomes at least 5.0 at the eye. Rule-of-thumb: **Any light that is admitted to the observer's eyes MUST have the IR/UV radiation removed and be at a safe level of intensity.** For white light photography consider inserting a supplementary filter before the detector, as outlined in Chapter 3, that enhances the feature of interest. Green is excellent for photographing granulation, blue for faculae, and red for the penumbral detail of a sunspot.

The detector (film or chip) should be suitable for what you are seeking to photograph. For instance, a high-speed, large-grain film for whole disc white light photography would be an inappropriate choice. Why? Because of image brightness the additional light sensitivity is not necessary, and the large photographic grain results in a loss of resolution. Slow speed, fine grain film, or an electronic sensor with finely sized pixels is desirable. Chromospheric photography especially might be done with a detector that has good sensitivity in the region of the spectrum being imaged.

Surprisingly, a detector that produces color images is not always the best choice for solar photography, although it is sometimes the only choice available. The reality is that the Sun in white light is generally filtered to exclude all but a broadband of color highlighting a particular feature under observation. Even an unfiltered view with a projection telescope shows the white light Sun to be composed of only the weakest hues; rarely is color evident. The monochromatic solar view is strictly limited to one color, often a very narrow slice of pure red or blue light.

A monochrome digital camera producing a grayscale image, or the film camera shooting black and white, is preferable to a camera doing color photography. The exception, of course, is during totality of a solar eclipse of the Sun. Both film and digital format grayscale images tend to be sharper than those recorded in color. With a truly monochrome digital camera this is because all the pixels of the sensor are used to produce the grayscale image. RGB digital photography necessitates that certain pixels are assigned to record specific colors in the RGB scheme, resulting in a loss in monochrome efficiency. Don't give up hope, though; a camera producing an RGB image can be very effective when the color channel capturing the best detail is used during the image processing stage. For aesthetic purposes color, when desired, is nine times out of ten added later, during photographic printing or as one of the image processing steps. Slow speed films have tight grain and likewise high-quality file settings with a digital camera produce the finest pixel size. Either of these conditions is preferred for superior images capturing fine detail.

A photographer has control over many aspects of the process: the quality of the telescopic system, any filter in use, shutter speed, the detector. The most critical factor, however, is the one over which he has little or no control, the atmospheric seeing. By being cautious with an observing site, avoiding certain problematic situations, and carefully studying local seeing patterns, one can have a limited amount of influence, but ultimately the observer is always at the mercy of the air.

From the amateur's standpoint there are two directions for attacking the visibility dilemma. One is "selective photography," which means making attempts only during moments of especially fine seeing. For the film user, from a cost perspective, this is the practical way to go. Even if tripping the shutter only when the seeing appears best, you will find that a majority of film frames will be discarded as you sort through the developed negatives. Timing an exposure is critical. But the odds of capturing at least several sharp photos are dramatically improved by photographing when the sky settles even briefly. The other method we will call "random photography." This entails shooting many images at the photographer's discretion and later sorting out the good from the bad. Inevitably a number of superior images will be captured using this method. This plan is especially suited for a digital imager who can acquire literally hundreds of photos during an observing session.

My own digital imaging procedure is a combination of these two methods, monitoring the seeing conditions and shooting a burst of many images when I see the sky settle. This technique decreases the large number of poor-quality frames I would normally discard, and increases my chances of capturing an especially sharp photo. Whether you select a film or digital media for photography, always shoot as many pictures as possible when conditions permit. Review the captured images at a later time, save the better ones, and toss out the poorest pictures. Lastly, archive only the best for study. These steps are followed by all consistently successful astrophotographers; rarely is a sharp solar photo culled from only a handful of images.

There are three basic configurations used for setting up a telescope and camera system. These are the direct objective, also called prime focus, the afocal, and the projection method. The equipment on hand and what is to be photographed will determine the configuration you select.

Direct objective photography is the simplest of the three, requiring only a camera with a removable lens and an adapter for fitting the camera to the telescope. This method provides the widest field of view, and depending on the size of your detector and the focal length of your telescope, is usually the method chosen for whole disc photography of the Sun. The telescope's eyepiece is removed and replaced by the lensless camera so that the virtual image created by the objective forms in the film plane or on the chip. The telescope's objective in this configuration becomes the camera's lens, in essence a very long telephoto lens.

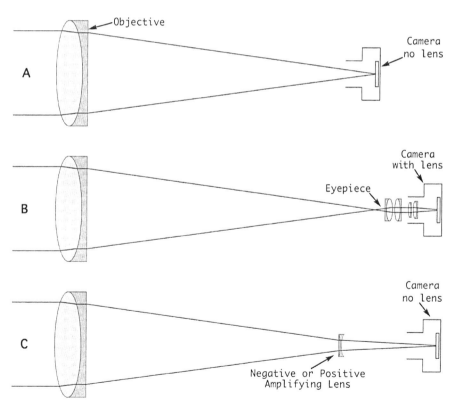

Figure 8.1. Three photography configurations. **A** is direct objective or prime focus method; **B** the afocal method, uses a camera with lens behind the eyepiece; and **C** is the projection method, with either a positive or negative lens used for enlarging the image formed by the objective.

A tiny detector or a long focal length fills the frame with a large disc of the Sun. Of course, too long a focal length for the size of the detector is not practical for capturing the whole solar disc. The frame of standard 35 mm film has on its shorter side a width of 24 mm. A digital camera may have a chip comparable in size to the 35 mm film frame or any of several smaller dimensions down to several millimeters square. Obviously, some digital cameras will not be well suited for this method of whole disc imaging. To compute the approximate size of the Sun in the focal plane of a telescope, the following formula from Chapter 3 can be used:

Focal length of telescope (mm) × .009 = solar diameter (mm).

Don't attempt to fill the frame completely, but allow about 5–10% of the disc diameter for the sky around the Sun to permit some freedom when composing a photo.

Some photographers have developed a work-around for photographing the whole disc with a smaller size chip. By photographing quadrants of the solar disc and then assembling them on the computer, a final picture of the complete Earth-facing hemisphere is possible. This is much additional work, but for the observer with a limited size sensor, it does provide a means of whole disc imaging.

One distinct advantage of prime focus photography when compared to the other methods is its fewer required optical elements. Having the least number of components, potential light scatter and wave front error is minimized, increasing the contrast and fine detail available to photograph. A well-made refractor or Newtonian is excellent for this type of photography.

One telescope design, the common Schmidt-Cassegrain, suffers from curvature of field, making it difficult to obtain an accurate focus across the entire solar disc with a flat plane film camera. Called Petzval curvature, the easiest solution is to stop the telescope down to about 100 mm (f/20) aperture, reducing the resolution but extending the depth of focus. Even stopped down this telescope will deliver its expected atmospheric resolution about 99% of the time.

Not every camera has a removable lens, so direct objective photography may be impossible. Many recent digital models have a fixed but focusable lens. For the astrophotographer having this type of camera, an imaging technique, called afocal photography, is used. This means that a telescope is focused visually, and the camera with its lens set at infinity is positioned directly behind the eyepiece. Light emerges from the focused eyepiece in parallel rays, like that from a distant object. The camera lens then brings these rays to a focus at the image plane. Even a simple point-and-shoot camera can be used in this manner. Mounting the camera to the telescope is done with a coupler of one type or another. Adjustable brackets attached to the telescope that utilize a ¼-20 tripod thread to hold the camera in place can be purchased or made in the home workshop. A commercial eyepiece adapter fitted between the camera lens and eyepiece, then inserted into the focuser, is a preferred option. Some manufacturers even provide special eyepieces that thread directly to the camera lens. You might like a unit that allows you to adjust the position of an eyepiece relative to the camera lens; that way you're certain the exit pupil of the eyepiece forms near the iris diaphragm of the camera. This position insures that vignetting (edge darkening) will be at a minimum in the camera. Scattered light can also be reduced in the camera with such an adapter, with the space between the eyepiece and camera lens sealed.

In order to calculate image scale with an afocal setup, it's necessary to determine the effective focal length (E.F.L.) of the telescope/camera system. The E.F.L. of a telescope is the total focal length of a system employing amplifying or reducing optics. With direct objective photography, the focal length of the telescope is a known quantity that can be used to calculate the linear size of 1 s of arc at the image plane from the formula:

$$\text{Focal length (mm)}/206{,}265 = \text{mm/arc second.}$$

For example, a 125 mm f/18 refractor has a focal length of 2250 mm (125 mm $\times$ 18). Insert 2250 mm into the above equation, and the resulting quotient tells us that at the focus of this telescope, the scale is 0.01 mm/second of arc. The total focal length *or* E.F.L. of an afocal telescope/camera system is found using this formula:

$$\text{E.F.L. (mm)} = \text{magnification} \times \text{camera lens focal length (mm).}$$

Let us again use the 125 mm f/18 refractor as an example, along with a 25 mm focal length eyepiece and a Nikon CoolPix 990 digital camera. The telescope with the 25 mm eyepiece provides a magnification of 90-power (focal length of telescope/focal length of eyepiece). The Nikon CP990 camera has an optical zoom lens with a focal length variable from 8 to 24 mm. When the camera is used in the normal lens position of 8 mm, the effective focal length of this telescope/eyepiece/camera combination is 720 mm (90 $\times$ 8 mm). When the camera lens is extended out to a maximum optical zoom of 24 mm the E.F.L. becomes 2160 mm (90 $\times$ 24 mm). The now known focal length of the system can then be inserted into the proceeding formula to obtain the image plane scale. Obviously, it can be seen that by selecting a shorter focal length eyepiece and/or a longer focal length camera lens, a greater image scale is readily attainable.

As for choosing an eyepiece, a high-end product is generally a better investment. A top-quality eyepiece produces a flatter field, tack sharp imagery, and better eye relief characteristics. A medium to slightly long focal length eyepiece (15–32 mm) performs well with a camera having a small lens.

Although afocal photography works well for capturing images, a camera's lens elements make the system a bit inefficient with light, and with a low-end camera, prone to aberration problems. Direct objective photography is the most efficient, but unless a long focal length is used an active region will be imaged small and not well resolved. Afocal photography can provide the appropriate scale for imaging detail, but the most efficient means is to enlarge the Sun with either a positive or negative lens, without light passing through the additional camera lens. This is called projection photography.

If you've tried solar projection as described in Chapter 3 for observing the white light Sun, you've seen the projection method in action. Substitute a detector for the projection screen, add solar filtration, and you have a basic layout for projection photography of the Sun. A positive lens (usually a well-made eyepiece), a negative lens (teleconverter, Barlow lens), or a combination of negative and positive lenses such as with a telecentric or Powermate^{TM} amplifier, is possible. Setups vary from one photographer to another, but the goal in each instance is to enlarge the primary image to a size that permits high-definition photography, often 3$\times$ to 10$\times$.

What about positive or negative lens projection? Each technique has its advantages and has been used successfully by experienced photographers of the Sun. An eyepiece of a suitable focal length is readily available in every observer's bag of accessories, making positive projection the first choice of an aspiring imager. The orthoscopic, plössl or other well-corrected eyepiece is recommended for positive projection. A few photographers have even culled the surplus optics from a used microfilm camera to serve as a projection lens. The distance that the eyepiece moves from its normal afocal position (focused for the eye) to that of near infinity magnification is but its own focal length. This short distance makes telescope compatibility for projection photography easy with standard focusers. The spacing of the camera and the eyepiece is the controlling factor for magnification, and additional units of power can be added with extension tubes equal in length to the focal length of the eyepiece.

In spite of all the convenience of positive lens projection, it is difficult to use an eyepiece for lower magnification work. As less and less distance is applied between the detector and an eyepiece, its small field lens begins to restrict light at the image edge. A more efficient alternative to projection with a positive lens is projection with a larger (30+ mm) diameter negative lens. The spacing will be similar to that of the positive lens, but a negative lens is positioned inside the cone of light from the objective, whereas a positive lens is on the outside. This makes for an even greater compact arrangement of the telescope/camera system, another advantage for the negative lens projectionist.

Computation of several elements is necessary for determining spacing for a desired E.F.L. Begin by deciding how large an image you want on the detector. If photography at the theoretical resolving power of the telescope is your intention, then the Nyquist theorem and the formula utilizing focal length and resolution as presented earlier must be considered. Should your interest be capturing a specific angular field of view, figure the focal length for the linear size of 1 s of arc at the image plane. Then take into consideration the dimensions of the detector; too small of a detector may not cover the field with the given focal length. Once the desired total focal length of the projection system is known, compute the projection magnification (PM) to obtain the E.F.L. for either a positive or negative lens with the following formula:

$$\text{E.F.L./telescope focal length} = PM$$

Now that it is known how many times ($\times$) the primary image must be enlarged to achieve the desired effective focal length, the separation of the projection lens and the detector can be calculated from the following:

$$\text{(positive lens) separation} = (PM + 1) \times \text{focal length of projection lens}$$

$$\text{(negative lens) separation} = (PM - 1) \times \text{focal length of projection lens}$$

Optimized projection photography demands that the path taken by the light be free of reflections, contributing to a scattering and general lowering of image contrast. The interior of adapters, extension tubes, etc., must be flat black. Cut threads or glare stops on the inside of an extension tube is a superior method of controlling stray light.

Here are some key points for high-definition solar photography:

1. Assemble a system that provides detailed arc-second resolution.
2. Choose a detector/filter combination that enhances the feature under study.
3. Focus often and accurately.
4. Use a shutter speed of 1/125 s or faster to freeze atmospheric seeing.

5. Monitor the seeing and photograph only when the view is sharpest.
6. If possible use a monochrome digital camera or b/w film; color can be added later.
7. Shoot many pictures, and throw most of them away.

Film as the Recording Medium

The film photographer likely will develop a technique of selective photography, making an exposure only during the fleeting moments of good seeing. The point behind this technique is to increase the chance of capturing a useful image, in spite of less than ideal seeing. This method is administered by monitoring the Sun through the same optics used to do the photography. With the same goal in mind several dedicated amateurs have constructed beam splitters that send a percentage of the Sun's light to either an eyepiece for direct viewing or to a video camera for remote viewing before it enters the film camera. The least complicated method, however, is to insert a clear focusing screen in a 35 mm single lens reflex (SLR) camera. Imprinted on the clear screen of camera should be a crosshair; when the crosshair and the Sun as seen through the viewfinder appear sharp, the system is in focus. This is more accurate than trying to focus on a coarse ground glass screen.

Other than focusing, another advantage of using a clear screen is the simulation of a visual observation. Peculiarities in the seeing conditions are apparent in the camera's viewfinder. The camera shutter is to be opened when the seeing is best. Since good seeing can be fleeting, a bit of practice and anticipation is needed to reduce the observer's reaction time when tripping the shutter. Expert timing comes only from practice, and it may take several rolls of film passing through a camera for an individual to master the technique. To suppress shaking of the camera/telescope system, try using an air-activated cable release to trip the shutter.

A camera with a low-vibration shutter system is mandatory, or blurring from mirror slap and/or shutter recoil will ruin the photos. Two older model SLR cameras are well known to have a minimum of internal vibrations, the Olympus OM-1 and the Miranda Laborec. The used camera market is the place to locate these models, the earliest versions of the Laborec being sold under the Mirax label. Useful accessories for these cameras include a set of interchangeable viewing screens and a variable focus magnifier. In Figure 8.2 a Laborec is pictured, a vintage 1960s model. The viewing lens on top of the camera is really a magnifier that works just like a telescope's eyepiece. Unfortunately, the earliest models of the Laborec have a limited shutter range, with a top speed of 1/125 s, just barely fast enough for solar photography. Nevertheless, for film work either the Olympus or Miranda camera comes highly recommended.

A bit of experimentation will be required at first to find the proper exposure for a telescope/filter/film combination. Freezing the seeing conditions takes a shutter speed of at least 1/125 s, and the photographer will want to bracket around an ideal setting in order to compensate for exposure variables. Begin by taking detailed notes of filters, films, and exposure times during a photography session. Compare the notes with your results after the film is developed, and use the best settings as the basis for future sessions. Factors that can alter exposure will include film speed,

Figure 8.2. Mirax Laborec camera for direct objective photography.

developing time and temperature, filter factors, transparency of the sky, and altitude of the Sun above the horizon.

With the rapidly changing photographic scene it is risky to suggest any specific film in the event it is removed from the shelf before this book is in your hands. Selection is best limited to a black and white emulsion having a slow ISO speed and fine grain. For H-alpha photography, a b/w film with extended sensitivity in the red region of the spectrum is preferred. Solar features are usually of low contrast, so film processing is usually left to the photographer with either a contrast-boosting developer or a doubling of the development time in a general-purpose developer. The key to success when working with film is consistency; once a successful procedure is established, stick to it, tweaking it only when necessary. Too much experimentation wastes film, becomes costly, and more importantly, leads to lost photographic opportunities.

Speaking of missed opportunities, you should be aware of some stumbling blocks that are encountered from time to time when doing astrophotography. For example, take care when loading the film into the camera, making certain the leader of the film is properly placed on the take-up spool. Be sure you have rewound the film completely into the canister before opening the back of the camera. And lastly, if the shutter is *not* linked to the film advance mechanism, always remember to wind the film after every picture. Many make those mistakes and a few more along the way. The most disheartening for me was discovering that a film leader had been improperly positioned on the take-up spool *after* a transit of Mercury was complete. Not a single picture was recorded, since the film failed to advance between shots.

In order to document a photo, notes should be taken that record what was photographed, when the photo was obtained, the exposure time, seeing conditions, and the equipment used. For example, a photo of an active region has little scientific value if the time the picture was taken is unknown. This is not unique to filmwork; a digital image must also be properly documented if the photo is to have any value beyond pure aesthetics.

Once the negatives have been processed and dried, the task of selection begins. A light box is useful for illuminating the negatives for inspection with a loupe or magnifying glass. First number the frames consecutively with a fine-tipped negative marking pen, matching the negative sequence with the notes you've taken at the telescope. Next cut the film into strips of six exposures and then inspect them with a magnifier for clarity and detail. Sufficiently sharp frames should be identified and the negatives stored in glassine sleeves until they are either photographically printed or scanned for digital manipulation on a computer. Scanning film gives the photographer a larger toolkit in the printing/publishing process; adjustment to brightness/contrast, sharpening, or the histogram are a few of the possibilities.

To the dismay of some, the digital camera is gradually supplanting the film market. We would encourage any novice solar astronomer just developing an interest in imaging to carefully consider the options. If you are already well versed in film photography and expect a small learning curve, then that past experience can be aptly applied to solar film photography. The inexperienced will probably be better off "cutting their teeth" with a digital camera.

Digital Cameras

The digital camera, or "digicam," market is full of products boasting all the latest bells and whistles. If you doubt this, just take a trip to a nearby variety store and see what is available in the camera department. There are models that include auto-focus, auto-flash, optical and digital zoom, a memory storage disc or card, movie-making capability, in-camera software image manipulation, and the list goes on and on. At first, this can seem overwhelming and somewhat frightening. If you have learned the ropes on film and are perfectly comfortable continuing down that path, you can do that. But some digicams have features that could make a solar photography hobby easier and more reliable. Because digital cameras do not use film, no extra cost is incurred for the film or processing, an immediate bonus. The memory hardware of a digital camera is reusable, and a typical memory card can store hundreds of photos. Because that quantity is important for beating atmospheric seeing in solar photography, this, too, is a plus.

Several of the more adaptable digicams are from the early Nikon CoolPix series (880/990/995/4500), each a high-end point and shoot non-SLR camera. Examples of photography accomplished with these marvelous little gems are found sprinkled throughout newer astronomy books and magazines. My personal introduction to digital astrophotography has been with the Nikon CP990; the discussion that follows is based on my experience with that camera (Figure 8.3).

The Nikon CP990 has a fixed lens attached to a telescope in the afocal position. Critical for superior performance of a digicam when used for afocal photography is a well-made camera lens. Remember that the more optical elements inserted into a light path, the greater the opportunity to create aberrations. The CP990 has a

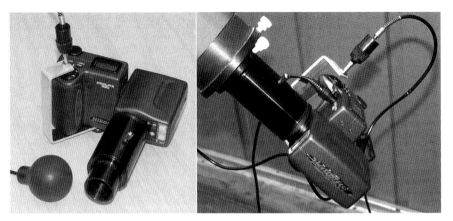

Figure 8.3. Nikon Coolpix 990 digital camera setup for afocal photography.

Nikkor brand lens, known for its exceptional quality and top-notch results. With the addition of a 28 mm T-thread coupler, my previously used extension tubes and film camera adapters were suitable for CP990 afocal photography. The T-thread coupler provides a connection between the lens of the camera and an eyepiece holder. The eyepiece is placed in the holder, locked into position with a knurled screw, and attached to the camera via the T-thread coupler. This package is inserted into the telescope's focuser.

The eyepiece I use is often of an orthoscopic or plössl design with an 18–25 mm focal length. For greater magnification, an achromatic 2.4× Barlow lens is inserted in front of the eyepiece. Many consumer digital cameras, like the CP990, have a built-in zoom feature. The zoom provides a wonderful means of increasing the E.F.L. of the telescope/camera system. Warning: utilize only the optical zoom and not a digital zoom, which only enlarges the central portion of the image to fill the frame. Unlike optical zoom, a digital zoom does not increase the amount of detail visible. It only increases the size of the image, making it somewhat more grainy appearing.

Battery drain can quickly zap the life out of any digital camera. An astronomer can carry a pocket full of rechargeable batteries or make use of the alternatives. Nikon cameras have an accessory AC/DC adapter for use as an external power source. Hardly feasible at a remote observing site, the adapter is practical anywhere standard house current is available. Purchasing a power adapter can be one of your wisest investments.

Nikon also makes available a remote release cord intended for activating the shutter so a photographer does not have to touch the camera. This can be a useful accessory, but you might want to use instead an air-activated cable release. Attached to the CP990 at the tripod socket on the underside of the camera, an aluminum bar may be bent so that the plunger of the air release presses on the shutter button when the bulb is squeezed. If the camera is in a continuous shooting mode, photos are taken non-stop until the bulb (shutter button) is released. The air bulb is preferable to the electronic release because the delay time of taking a picture is less with the regular shutter button. Good seeing conditions are brief, and it is important to have the shutter open while the sky is steady.

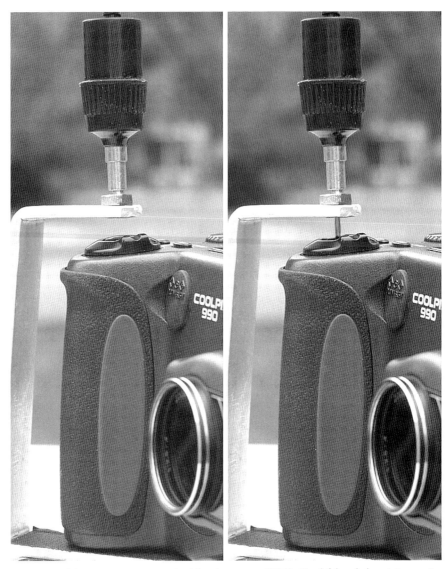

Figure 8.4. Adapting an air-activated cable release to a CP990. The *left hand photo* is in standby mode, and the *right hand* is with the plunger depressing the shutter button.

My camera is set to photograph in the manual exposure mode with the aperture wide open and the shutter speed adjusted accordingly for proper exposure. It is not unusual for me to bracket exposures around a suggested setting determined by the camera's light meter. I use a variety of broadband filters when shooting the white light Sun, and the exposure factor of each filter is different. The camera's meter is a quick and accurate way of determining the approximate exposure. Sensitivity of the detector is set to ISO 100, using the black and white recording mode with a maximum image size of 2048 × 1536 pixels. The format selected for saving a picture produces a mildly compressed JPEG file. This setting decreases the recycle time of the camera from what would be necessary if I were saving a TIFF file. The quality of a JPEG is hardly compromised when using this compression, and the benefit from the reduced file size outweighs any gains experienced from saving an original photo in TIFF. The camera's lens is set to infinity and the continuous photography function activated.

One unique software option with the CP990 is a function called best-shot selection (BSS) that automatically compares up to a group of ten photos and saves the sharpest to the camera's memory. I use this on occasion to reduce the number of throwaway images that would normally accumulate during a photography session. Other characteristics such as contrast, brightness, and sharpening are more subjective, and I don't normally offer recommendations in those areas.

Finding accurate focus and monitoring the seeing conditions, particularly during the daytime, can be nearly impossible on the tiny LCD of most cameras. To combat this many photographers feed the signal from the camera to a high-resolution TV monitor. When it comes to off-camera viewing, check out the 210 mm (9-inch) black and white security monitor available from Radio Shack. Plugged into the video-out port of a camera, this monitor is powered from a standard AC wall outlet. Finding focus is literally a snap; aligning the camera to celestial east-west and framing photos is easy, too, with the magnified on-screen image of the TV monitor.

It is extremely important to control glare and reflections from ambient daylight on the large screen, so a hood is fashioned from black foamboard and attached to its casing with Velcro. A peephole cut just above center on the screen allows a photographer to view the screen in the shaded environment. Other possibilities to enclose a monitor include a large cardboard box, a photographer's black cloth or towel draped over the monitor and observer, or consider moving the monitor indoors to a control room while slewing the telescope remotely.

Occasionally, things go wrong when using a digicam for afocal photography. Vignetting, or the falling-off of light at the edges of an image, is common and can be the result of several mistakes. If a camera is not centered or mounted perpendicular to the optical axis, some corners of the image will take on a gradient appearance, in extreme cases a solid black. Most likely the cause is a mismatch of the telescope's exit pupil and the entrance pupil of the camera. For best results, the telescope and eyepiece exit pupil should be equal to the camera entrance pupil.[1] Or, it could be that the camera is positioned too far from the eyepiece; if this is the case, readjust the camera relative to the eyepiece so that the iris of the camera lens is located at the eyepiece's exit pupil.

Random pixels that are unnaturally saturated create a speckled effect in a photo, called noise. One possible solution is to decrease the ISO setting of the sensor and increase the exposure time. Also, noise can result from a camera that is literally hot, particularly one located in bright sunlight for a period of time. In Figure 3.8, amateur astronomer Art Whipple uses a space blanket type material to shield his video camera from direct sunlight. If heat generated from the digicam itself is the culprit, turn off the camera for a regular cool-down period when necessary. Another method to artificially curb hot pixels in a photo is to layer or composite several images during processing that have been obtained close in time. The several images should have the frames shifted slightly so the hot pixels in each are not layering over one another. This stacking technique effectively fills in the missing data from the primary image.

A typical imaging run with the CP990 follows this sequence. With the monitor connected to the camera through the video-out port, the digicam package (camera/adapter/eyepiece) is inserted into the telescope focuser. The monitor and camera are powered up, and the Sun is located in the telescope by watching through the shade's peephole on the security monitor. Rough focus is made, and the camera body rotated in the focusing unit until the east-west line of celestial declination is parallel with the long side of the image frame. A rough sketch of the region to be imaged is now drawn with notes made of the celestial directions (north and east) for later reference when processing the images. Using the slow-motion controls of the telescope, the area to be imaged is centered in the monitor, and fine focusing is performed. A meter reading is taken to determine the approximate exposure time needed. A test exposure is taken, and the Universal Time is noted so interpolation can be used to find the time of succeeding images from their file creation record. Adjustment is made to the shutter speed.

Now the region to be imaged is patiently watched on the off-camera monitor; focus is adjusted as needed, and the telescope is slewed to keep the region centered in the field. Whenever the view becomes steady, the shutter is tripped for a burst of several images, followed by the camera's BSS software selecting and saving the sharpest of a burst of pictures. This continues until a large sampling of images is taken or if a particular event has ended. When imaging is complete, the camera and monitor are powered down and the digicam package removed from the focuser.

White light and monochromatic solar images captured with a consumer digital camera provide the means to rapidly see spectacular results. The obvious advantages of this system are realized in the cost savings and convenience of obtaining many images when compared to film. The disadvantage is that a fixed camera lens necessitates afocal photography, a technique that subjects the optical system to possible aberrations from the eyepiece and camera lens.

DSLR Cameras

Take a digital camera and merge it with a SLR and the result is a digital single-lens reflex camera, or DSLR. This camera design appeared in the marketplace not long after the digicam revolution began. The advantages of using a DSLR are the ability to remove the camera lens and to view directly through the optics taking the picture. To the photographer an interchangeable lens means freedom to select direct objective or projection photography. Ordinarily the sensor in a DSLR will be larger than that of a digicam, allowing a wider field to be imaged. Many DSLR cameras also have a video-out port that permits the use of an external monitor for focusing, framing, and studying seeing conditions. A few models may have a built-in magnification function that allows accurate focusing directly on the tiny viewing screen.

Most digital cameras are engineered to create a color photograph through a technique called Bayer Masking, in which a color filter array is built into the CCD/CMOS imaging chip in a replicated pattern of red, green, and blue. Green, because of its increased sensitivity to our eyes, is given dominance in the RGB pattern. The green pixels are also selected for imaging when a black and white conversion is done in the camera. Fully half of the pixels in a chip are green sensitive; the remaining red and blue are 25% each.

For white light solar photography, the primary filtered incoming signal is normally a broadband containing to some degree these three colors. Fortunately, many observers imaging in white light use a narrow band green supplementary filter to accentuate faculae and solar granulation, which makes use of the majority of pixels in the RGB sensor. On the other hand, a camera used for H-alpha or Ca-K photography loses out on pixels because the transmission of the solar filter is so restrictive that pixels outside either the red or blue are barely affected. A monochromatic photographer, in theory, utilizes only 25% of a digital camera's pixels. Some pixel arrays suffer from a defect called leakage.

Leakage happens when light from one channel, either the R, G, or B, spills onto adjacent pixels and creates a false color image. A crafty photographer can use this to his or her advantage when processing files. We'll explain this later in the chapter. The best option for a H-alpha or Ca-K photographer to shoot is with a monochromatic (B/W) camera, which uses all the pixels to form an image.

Lastly, it ought to be noted that for an accurate color match with general daytime photography, sensors have an infrared (IR) blocking filter positioned by the manufacturer over them. This IR filter doesn't typically have a sharp transmission cut-off that extends into the visible light. Some of the red light including H-alpha may therefore suffer a reduction in transmission to the chip. High-end custom astrophotography cameras will have the IR filter removed. As illustrated by Figure 8.5, regardless of these limitations, a commercial DSLR camera does a fantastic job of producing images of the Sun in monochromatic as well as white light situations.

Two makers of the DSLR camera, Canon and Nikon, have been quite popular with astrophotographers. At the time of this writing the Canon D60 or 10D and the Nikon D-series of cameras are suitable for solar imaging. Because the line of DSLR cameras is constantly in a state of improvement, any further recommendation of a specific model will become dated with the passage of time. Rather, you should explore the line of current models and discover which address the requirements of solar photography. The modern DSLR can be used for spectacular images of the Sun, Moon, and deep-sky objects, while doubling superbly for general daytime photography.

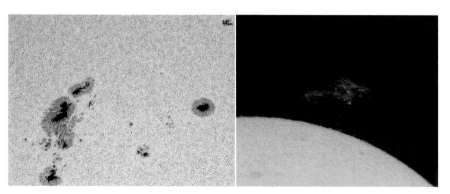

Figure 8.5. White light and H-alpha images taken with a Canon D-60 digital single lens reflex camera by Eric Roel. The telescope is a 150 mm f/12 APO triplet refractor.

Webcams and Digital Imagers

Enter onto the scene the webcam, an imaging device that started out as a black and white video conference camera mounted atop a home or office computer. Several enterprising individuals began cannibalizing the webcam to create a lightweight, low resolution, digital video camera for imaging the Moon and planets at a fraction of the cost of a dedicated astronomical CCD camera. Popular models included the Philips Vesta and the Logitech Connectix. My own introduction to webcam imaging was with a Connectix camera that I had disassembled and rebuilt into an aluminum canister style body. Amazingly the entire camera weighed only a few ounces, no more than a premium eyepiece. The homemade imager slipped into the telescope's focuser and was powered by the Apple computer I used to operate the camera's image acquisition software.

Focusing and framing the image or monitoring the seeing could all be accomplished directly from the computer screen, which was shaded from the Sun by a cardboard box having a hole cut into its front. Since that humble beginning, higher quality conferencing webcams have come along, supplying greater resolution, better bit-depth (ability to distinguish differences in tone), and even a full-color image. Astronomical equipment manufacturers have also developed a product similar to the home-assembled webcam, a digital imager that attaches to the telescope and operates through a laptop computer to shoot video clips or single-frame pictures.

The webcam and commercial digital imager is a fantastic tool for introduction to CCD astrophotography. And what they do better than other imaging options is create videos. The live image display with essentially no refresh time is a spectacular bonus. Several observers can be positioned around a laptop monitor to witness an event simultaneously.[1] Where these cameras have really excelled in the amateur ranks is in the field of planetary photography. The ability to capture literally hundreds or thousands of individual frames and then, with special editing software, pick out only the sharpest pictures and combine them into a single image has been the key to success for the Solar System imager.

As we will see, besides planetary work, a webcam is also suitable for certain types of solar photography. The chip of a webcam or digital imager usually is 640 × 480 pixels; some cameras with larger chips have a 1280 × 1024 pixel resolution. Regardless, the physical size of a chip in a webcam is limited to approximately 8–12 mm square. Although not an issue for a planetary photographer, whose subject is relatively tiny and fits easily inside a chip, the Sun's large disc is hardly able to squeeze within these physical dimensions. Producing a high-resolution whole disc photo of the Sun with a webcam requires the stitching of numerous images into a single panoramic photo. It is a bit like assembling a puzzle that has many square pieces. It can be done, but it's a lot of work. It should be obvious by now that the value of a webcam for a solar astronomer is high-resolution imaging of small regions of the Sun.

Solar observer Steve Rismiller uses a Phillips Toucam Pro Webcam with a Vixen 102ED telescope to capture white light and monochromatic images of the Sun (Figures 8.6 and 8.7). Rismiller writes, "The Toucam has the lens removed, and is fitted with a 1.25-inch diameter nosepiece. A USB cable is used to power and transmit the AVI movie files to my laptop. I don't like sitting out in the blazing Sun to observe, so I have lengthened the USB cable, the connection to the electric

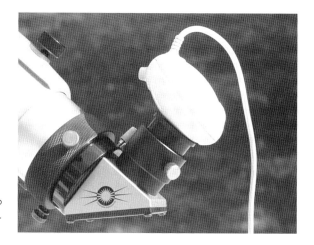

Figure 8.6. Phillips Toucam Pro Webcam mounted for imaging. Steve Rismiller.

focusing unit, and the mounting's slow-motion control cable so I can sit in our 'sun-room' in the shade. I can control the scope from 15 feet away, under the comfort of a ceiling fan. Another benefit from indoor observing is that the computer monitor is easily seen".

To capture images, Rismiller uses the software, K3CCD Tools and WcCtrl-WebCam Control Utility. A camera can be used with the pre-packaged software that comes with it, but specialty software packages written specifically for astronomical use are convenient and make image acquisition easier. The Toucam captures in a 640 × 480 pixel format. Accurate focus is obtained with a digital magnification of the on-screen image and is carefully adjusted by eye. Inspecting the histogram of the image sets exposures. Correct exposure is found when the histogram displays pixels distributed across the brightness range and does not

Figure 8.7. Toucam closeup. Steve Rismiller.

show a "clipping" of either the highlight or shadow ends. Rismiller's technique is to set the K3CCD Tools video software to shoot continuously for 10 s or more, producing an AVI file containing around 145 separate images. A number of these raw AVI files are obtained during an observing session, with the best images later culled from each file using Registax software.

Ready-to-go commercial digital imagers are also available that can be adapted for solar imaging, as demonstrated by Howard Eskildsen in Figure 8.8. Eskildsen's camera is the Orion StarShoot II, with a 1280 × 1024 pixel sensor. Operating primarily out of the trunk of his car during weekday lunch breaks or in the driveway at home, Eskildsen has assembled a quick setup and takedown system for high-resolution solar imaging.

The key to Rismiller's and Eskildsen's photography technique is to capture a large quantity of images within a relatively short time period, later pulling out those images that are sharper than others and deleting the rest. Eskildsen normally captures images for 30–60 s at 15 frames per second. If time permits he will collect several of these files, in the common AVI format for future editing. Speaking of his technique Eskildsen says, "Following image acquisition, the editing process continues as the AVI frames are aligned, and stacked with Registax. Usually between 20 and 100 acceptably sharp frames are picked, optimized, and stacked from a 60-second file. The wavelet function in Registax is used to sharpen the stacked image. I try to maximize the sharpening without inducing artifacts, and often experiment with the settings until I find one that is eye-pleasing." Final processing of the stacked image is done with Photoshop to adjust the orientation with north up and east left. Further adjustments are made to the histogram using the levels function while brightness, contrast, and sharpening are mildly tweaked for aesthetics, making the image ready for presentation.

Figure 8.8. Solar imager Howard Eskildsen with his portable setup for digital photography. The opened-face box serves to shade the laptop from the direct light of the Sun.

Dedicated Astronomical Cameras

For the serious solar imager a variety of high-end dedicated cameras for astronomical photography are available. These cameras fall into two basic classes of image capture, single frame or video acquisition. Some models are capable of performing both tasks. The high-end products offer a sensor that is usually larger in size than that found in most commercial digicams or webcams. The pixel size (in microns) of some high-end cameras may actually be slightly larger than that of the lower-end products, but because of the greater width and height of the chip, a magnification of the primary focus image along with the larger field of view is possible.

A dedicated camera for the serious observer is usually monochrome, that is, the images produced are grayscale. Monochrome cameras make use of all the pixels in the array to form an image, unlike an RGB chip, which is dependant on a technique such as Bayer Masking to capture color. Color photography with a monochrome camera is done through a process called tri-color imaging in which separate photos are obtained with color separation filters and then digitally combined into a single photograph.

Effective white light or monochromatic solar imaging is really a grayscale process; color solar photography is only an aesthetic procedure performed through image processing. When a feature is reduced to what's seen in the eyepiece of a solar telescope, 99% of the variation found is in the intensity of the light reaching the eye. What does this mean? In H-alpha all the light is red, in Ca-K all the light is blue, in white light the hue is basically flat, generally biased by the transmission color of the primary and supplementary solar filters. Any color differences visible in a sunspot umbra will be extremely subtle and is within the realm of a visual observer. Shooting a digital solar image is therefore best accomplished with a monochrome camera because the engineering of a sensor to produce a color image reduces the effective resolution of the chip through techniques like Bayer Masking (Figure 8.9).

A high-end camera will have an improved capability to image variation in tone between black and white. The amount of gradation that can be registered in a

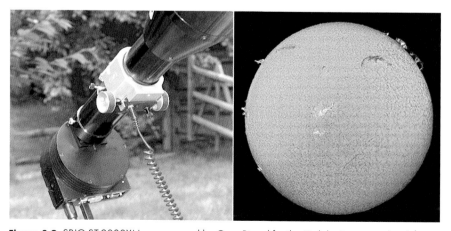

Figure 8.9. SBIG ST-2000XM camera used by Greg Piepol for the H-alpha image on the *right*.

single exposure is known as the sensor's bit-depth. This characteristic is referred to as bit-depth because each pixel is in reality a solid color or tone defined by a "bit" or number. Digital images are in essence a huge collection of numbers called computer code. With the simplest imaging system, only a number of 1 or 0 can be assigned to a tone, making it either a solid black or white. This situation is called a 1-bit file. A little more complex is the 2-bit system, in which there would be four possible tones: white, black, and two variations of gray between them. The computer code for a 2-bit file can contain the values 01, 11, 10, and 00.

The eye needs nearly 250 shades of tone from black to white to create a smooth appearing grayscale, or what is termed a continuous tone image. An 8-bit system allows enough combinations of ones and zeros in computer code to produce 256 different shades of black and white. You will find that most digital cameras use a sensor producing an 8-bit image in grayscale and a 24-bit in RGB mode (8-bit × 3 channels). There are instances, particularly with H-alpha imaging, when a photographer will want to record prominences at the limb as well a surface detail in a single exposure. It becomes impossible, however, to capture detail from both adequately because the correct exposure for the disc will be shorter than the correct exposure for prominences. It becomes an either-or situation. This characteristic, the ability to register detail having a great difference in brightness, is known as dynamic range. The greater the dynamic range the better, and while film tends to have the advantage in this area, developments in technology could allow a digital camera one day soon to exhibit remarkable advances in dynamic range. The solution for a solar photographer wishing to record disc and limb features is to combine two separate and correctly exposed images into a single photo.

An amateur astronomer will likely want to feed the signal from a camera directly to a portable laptop computer. It is important if you are contemplating the purchase of a camera to consider the compatibility of the two pieces of hardware. Issues to be addressed include: Will the memory allocation and processing speed of the laptop be suitable for this camera (more of either is better)? Is the operating system compatible with the camera's drivers and acquisition software? And what about the connectivity of the camera and the computer (newer cameras use a USB 2.0 or FireWire connection)?

Veteran solar imager Greg Piepol has had extensive experience with several types of high-end imagers (Figure 8.10). His Santa Barbara Instrument Group (SBIG) Model ST-2000XM camera features a high dynamic range, low noise, and a 2-megapixel image for large, detailed photography. Greg relays this information regarding his experience with a dedicated astronomical camera for solar imaging:

> The SBIG is not a live feed CCD; it requires about ten seconds to download the 2mb file. I first compose the shot (obtaining focus, position, and correct exposure) in the "focus" mode of the camera's control software. It usually takes four or five minutes to get it perfect. Each preliminary composition shot takes 15 seconds total (about 5 to capture and 10 to download). I make a change if necessary, shoot another test exposure, and wait 15 seconds. On and on the composing routine continues until it's perfect. Then I switch to the "grab" mode and capture a shot. I will do several grabs in a row. Wait a little while for the seeing to calm down, and do it again. I observe with my laptop from a small portable trailer dubbed "the Sunspot," drawing a curtain to keep the daylight out; focusing and slewing are done remotely. The scope is set up just outside the trailer.

Figure 8.10. The Luminera video camera can shoot at 200 fps, but lower frame rates improve the resolution. Greg Piepol.

Amateur video captures may also be done with a high-end solar system camera. Several makers include Lumenera Corporation, The Imaging Source, and Adirondack Astronomy. With a camera having the option to shoot video, consider the camera's frame-rate. The more frames per second a camera records, the greater the odds are of capturing a brief moment of steady daytime seeing. Experienced imagers say that 30–60 frames per second are excellent, although slower rates can be effective. Piepol notes that to collect a video clip containing about 100 individual frames with his Lumenera camera requires around 10 s of time. Lowering the resolution of the captured video permits operating at a maximum of 200 frames per second. But by using the full resolution, the fps is only limited to 15, just fast enough to capture moments of steady seeing. The lower the noise characteristics of a sensor, the smoother the image will appear to the eye. A high-gain setting increases sensitivity of the chip but also the noise level. The video-producing camera that compresses a file to save memory introduces a small amount of noise, creating lower quality output. The low or non-compression file would be the best option for saving single images or video clips. Connection to the telescope focuser with this type of camera is usually done with a threaded standard C-adapter, available from most camera suppliers.

Processing an Image

Not that long ago an amateur photographer might have shot a roll of film and then rushed off to his or her local camera store to have it developed and printed. Perhaps several days later the photographer could return, and the store would have the photos and negatives available for inspection.

Today, the story is far different. The digital camera has opened the floodgates to even a novice when it comes to homespun photography. An average photographer might download pictures directly from a digital camera to a desktop printer and within a few minutes have a hardcopy photo for study and sharing. With an option to adjust the overall appearance and scale of a photo more possibilities for the photographer are created. These alternatives are executed by downloading the original image files to a computer equipped with photo-editing software. File manipulation or editing embodies rotating and cropping an image or making adjustment to the histogram, contrast, or brightness. It is even possible to improve the sharpness, or decrease the amount of noise in an image. Detail can be brought out by putting dozens to hundreds of individual frames one on top of another with a technique called stacking. The computer allows us to stitch several adjacent images together to form a panoramic view, or we can add color where none existed before, to turn a drab black and white image into a vibrant, colorful one.

Always perform your editing on a copy of the image. Once a change has been saved to a file it is impossible to return to the original image. Take care, however, to avoid over-processing an image. Being zealous with editing functions can be carried to the point of creating artifacts or adding non-existent detail to an image. It is acceptable to enhance what is barely visible but sloppy to create non-existent features and unprincipled work if done intentionally.

The die-hard film enthusiast doesn't have to give up film in spite of the trend away from the darkroom. By embracing a "digital darkroom" the film photographer can have the best of both worlds. What does this mean? Negatives and transparencies can be imported onto the hardware of a computer, and then edited just like an image obtained directly from a digital camera. The digitized film image can be printed as hardcopy, similar to a conventional photographic print, or displayed electronically as a digital file.

Photographic film represents an analog technology. That is, photons are recorded in a direct or linear fashion through a chemical process. Digital photography captures photons and records them as binary data. The task for a film photographer is to convert the analog information to digital data by using a peripheral device called a scanner.

Because of reproduction limitations in the photographic process the best method of conversion is to scan the original negative or transparency rather than a second-generation print. Compared to the original film a hardcopy print has a reduced brightness range. There is information in the negative that will be lost through the printing process and is unrecoverable. Film scans that are high resolution (comparable to the output of a digital camera) are desired, with a minimum depth of 8 bits per channel. Because film images often have a greater dynamic range than is recordable in a single scan, it may be necessary to scan some features at one setting and the weaker appearing features at another, then combine the two images digitally to produce a single image.

File Format

File format is basically the arrangement of words and illustrations within a digital file. It can be described as "how a file is saved." Some image formats create a file that is huge, many megabytes in size, while other formats use compression to reduce file size. Sometimes a file will open only within a specific software application, while another format, for instance the JPEG (pronounced JAY-peg) is compatible with nearly all image applications and computer platforms. Of the many possible formats available for a still photograph, three are most often used for creating and editing images: the Raw, TIFF, or JPEG format. Each format has a distinct advantage and use in the imaging process.

Raw image files are in a manufacturer's proprietary format. The purpose of saving a file in a raw format is to record an image without internally created software effects tweaking the data from the camera's sensor. Raw files contain more information by several times than other file formats because they have practically no compression. Little or no compression means that little or no data has been discarded. Working with a raw file in the past was difficult, because not all photo-editing software could open it. The workaround for this situation was to convert the raw file to a different file format, one compatible with your editing software. Today, a number of photo editing programs are able to handle a raw file. Always make certain you have one of these software versions if you desire to use a proprietary format. Pixels in a raw file have a greater bit depth than the typical 8-bit file produced by other formats. Working with a 12- to 14-bit file means that the image will exhibit a finer gradation of tone and more detail.

While raw would seem to the purist as an ideal format in which to save images there are several disadvantages. File size can limit the number of images that are stored in a camera's memory. Remember that with solar photography, the more images the better. Not every camera necessarily saves an image in a raw format, making it impossible to even consider the format in some cases. The download time for a raw file is longer than other compressed formats, making it difficult to shoot images in rapid succession. And the proprietary formats are not standardized as yet. Photo-editing software today that works well with your raw format may not be available 5–10 years from now, not to mention the format itself. This condition will necessitate storage of the raw file in another format for any future use. As you can see, there are a number of trade-offs when using a proprietary image format, but if you are seeking a file with maximum data retention and minimal artifacts then the raw format is for you.

Tagged image file format, or TIFF (rhymes with IF), is the most popular format for editing black and white or color images on the computer. Practically all graphic manipulation programs and scanners support TIFF, giving the TIFF file a marked advantage over the raw file. Like a raw file, the TIFF also requires a lot of storage space. This means that the download time for a TIFF will also be substantial when compared to other compressed formats. Besides the compatibility of a TIFF to most editing programs, the ability to save data in what is called a loss-less format makes it a superb choice for archiving photos. Loss-less files are those that can be manipulated and re-saved without creating a loss of data. This removal of data ordinarily occurs during compression of the file, a condensing process intended for storage purposes. If desired, TIFF files can be compressed in a loss-less format using the optional LZW function (Figure 8.11).

The all-around format for a digital still image is the JPEG. An acronym for Joint Photographic Experts Group, the JPEG is a generic format recognized by virtually all computers and photo-editing software. The JPEG file is smaller than a raw or TIFF file, yet can exhibit high picture quality. Not all digital cameras save a picture as raw or TIFF, but all cameras do save images as JPEG. Internet web pages contain graphic images as JPEG, and a photo shared via e-mail with your friends is usually in the JPEG format.

Any format that reduces file size by discarding information through compression is referred to as lossy. A JPEG file falls within this category. The more compression and the smaller the file size becomes, the more its picture quality suffers correspondingly. An original image coming from the camera, particularly if it has had only a mild compression, looks very good, but each time a JPEG is

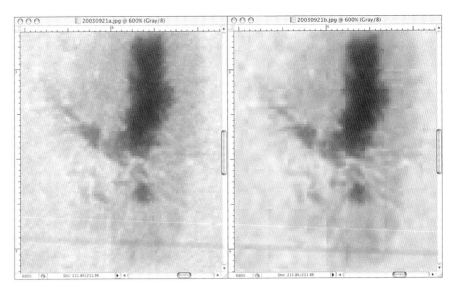

Figure 8.11. The result of editing and saving an image in a lossy format is evident in these 600% enlargements. The JPEG on the *left* was opened, manipulated, and resaved with a mild compression factor creating the *right side* copy. Close inspection reveals that data has been lost as the image becomes more pixilated. Higher compression loses still more data.

opened, modified, and re-saved, compression discards more information. It is a cumulative effect, enhanced more so when a high compression level is used. Choosing a lower compression setting when re-saving the file will minimize the loss.

The solution to data loss many photographers have seized upon is to conduct the entire image editing in a loss-less format, such as TIFF, and then save-out a JPEG file for use on the Internet or email. The edited TIFF file is archived for future generation of more JPEG copies. The amount of compression given a JPEG is determined by the intended use of an image. Graphics meant for building Internet web pages can be given high compression, because compromised quality may not be as important as download time. A patrol photograph for archiving with a national observing organization, however, needs to be of the highest quality, so a mild compression is in order.

To summarize, an original image obtained with your digital camera may be saved in either a raw, TIFF, or JPEG format. Raw and TIFF files are of higher quality, but file size is a stumbling block with those formats. A low compression JPEG is also acceptable as an original file, in fact a requirement if images are going to be obtained in rapid succession. Always edit a copy of an original image in a loss-less format, such as TIFF. Archive the edited loss-less image and output any files for distribution as JPEG, the compression factor set from low to high, depending on the intended use of the file.

Photo Editing Programs

Remember the days when editing a photo consisted of "dodging and burning" an enlargement during the printing process? Or to bring up a faint spot in the print, a finger was gingerly scrubbed in an area on the print while it was still in the developing tray, the friction warming the region and speeding-up the chemical reaction? A retouching pencil back then was a handy accessory for removing a dust speck or creating an artificially sharp border between fuzzy light and dark sections.

Today, with a keystroke or the click of a mouse, these manipulations and many others are possible when a digital image is "retouched." Most of the current photo-editing software programs are compatible with the several platforms on which computers operate. Most manipulation programs offer adjustments over brightness and cropping; other programs permit total control of images or a unique function that is not necessarily available elsewhere. It is not unusual for an advanced imager to perform some manipulations in one program and finish processing the image in another.

Taking particular software and presenting a step-by-step approach to the editing process will limit the usefulness of this book. This is for several reasons. First of all, visual aesthetics is truly a subjective topic. What one person finds pleasing to the eye another might not. Secondly, the "learn by doing" method is really the best. We will provide you with the basics, an indication of what can be done, and then turn you loose to find your own way. Besides, software programs inevitably provide several avenues when performing the same task, to allow for the differing thought processes of the various users. There are many books on the market that focus solely on manipulation, permitting greater instructional space than there is room for here.

And there's another issue. Revisions and updates to software are continually coming down the pipe. For instance, Adobe's Photoshop 3 was the latest version available when this author began working with digital images. Since then there have been at least six newer versions replacing that Photoshop program.

Some of the basics will be presented here, though, while leaving the door open for the digital imager to find his or her way through it. With the current software available, finding your way will be easy. High-end editing software helpful for solar photography includes Maxim DL, ImagesPlus, Adobe Photoshop, and the popular Registax.

Calibrating the Monitor

In the digital medium, the acronym WYSIWYG was popular for a time. "What You See Is What You Get," particularly with an image, meant just that – unless, of course, your monitor or other peripheral devices were out of calibration. Calibration is the adjustment made to a viewing or output device to obtain a uniform standard. It is the guarantee that the image seen on your computer screen will appear the same on another computer that also has been properly calibrated. If you output hardcopy prints you will want to calibrate the printer and monitor to ensure the final result matches what is seen on the screen. Most computers contain software that enables a user to make the necessary calibration adjustments to brightness, contrast, and color.

Although calibrating is a not a tedious job, there is no guarantee everyone's equipment will be up to the standard. To assist other observers with viewing solar images, you can embed a multi-step grayscale strip inside your final image to encourage an outside viewer to at least adjust his or her monitor's brightness so all the steps are seen clearly. Monitors that are set too dark will lose shadow detail,

Figure 8.12. Grayscale for quick screen calibration.

and the image will appear muddy; too bright, and the image becomes pale and washed out. Choose a strip that has a gradient with 17 steps that run from pure white to solid black. A similar grayscale can be downloaded from numerous web sites dedicated to calibration or you can make one using Photoshop in about 5 min (Figure 8.12).

Rotating and Cropping the Image

Proper orientation of a patrol type image of the Sun will have solar north up and the east limb toward the left. Some observers don't wish to go that far and are content having celestial north up with celestial east to the left, roughly matching the Sun's appearance in the sky. What's important is to establish a plan encompassing uniformity when orientating images. Using the techniques described in Chapter 4, the original image file will be aligned in the celestial cardinal directions. The daily value of P is used to adjust the Sun's tilt east or west, turning the solar northern pole vertical. You may also find it necessary to first "flip" the image horizontally or vertically to achieve this north and east orientation. This is why it is a good idea to make a sketch at the telescope, confirming the directions on the Sun. It may be several days before an image is processed and memory can fail.

The rotation tool of most photo-editing programs allows the image to be adjusted by fractional increments of a degree, giving very fine control over the final output. It is important to know the correct amount of rotation an image requires at the outset, because each additional nudge with the rotation tool may degrade the final image by interpolating calculations that throws away data. The fewer number of tweaks made with the rotation tool the better.

A cropping function works like a pair of scissors in the hand of the computer operator. An image that is too large can be trimmed smaller. Perhaps after rotating the image, it looks crooked, with white spaces peeking out from the canvas near the corners. The crop tool is used to trim the image to a square or rectangular shape, giving a neater appearance.

Some cropping software also incorporates a resampling option that may include scaling an image in height and width. Resampling happens anytime the quantity of pixels in an image changes and the dimensions of the image remain the same, or visa versa. There are two types of resampling: downsampling and upsampling. Taking an image that is 4-inches square at 72 dpi and changing it to 2-inches square also at 72 dpi illustrates downsampling. To maintain the same dpi resolution, information has to be discarded from the file, making it physically smaller. Whether the downsampling is acceptable or not depends on the intended use of the file. Does it matter if some information is lost? When creating a small graphic for use as a thumbnail on a web page, the answer would be no. If a file's intention were to be stitched with a dozen other images to create a high-resolution panoramic photo of a large sunspot group, the answer would be yes. In that case it is necessary to retain all possible information from the original file.

Upsampling is the reverse situation. Take a 2-inch square image at 72 dpi and create a new file that is 4-inches square at 72 dpi. Pixels have to be added to the image through a process called interpolation. The pixels are created mathematically, according to the characteristics of the surrounding pixels, and do not represent actual detail. From a scientific standpoint this is unacceptable because the editing process is now basically creating something from nothing. Avoid upsampling an image and use downsampling only if necessary to create a usable file matching the output requirement.

Histogram Adjustment

The histogram display is a guide that tells an imager the quantity of pixels present at various brightness levels in an image. Good grayscale images tend to include a wide range and quantity of pixels from black to white. A poor quality image will have less of a range. The result is a flat, muddy, and rather dull-looking picture. If the displayed pixels are spiked at either end of the histogram and fill up the pure white or black regions of the scale, the histogram is "clipped." Clipping commonly happens when an image has been over or underexposed. Adjustment to the exposure will correct this situation unless, of course, the bulk of the image is actually a pure white or black.

Solar features tend to be of low contrast; therefore, the histogram of some images may appear similar to what is seen in Figure 8.13. The display box of the left image is showing a brightness range of approximately 90 gray levels, revealing a flat and veiled photosphere. To improve this condition, the histogram of the right hand image has been stretched, using the function's adjustment sliders until over 250 levels are present. Now, the unique light bridge in the upper spot is evident, and the faculae surrounding the group are becoming increasingly visible. The comb-like histogram that results from stretching a narrowly compressed original

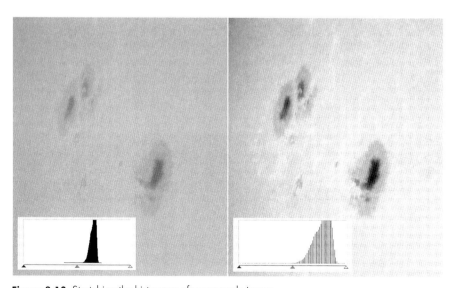

Figure 8.13. Stretching the histogram of a grayscale image.

indicates that gaps remain within the tonal range. The image is not really is a smooth gradient but is missing pixel data all across the brightness range. There is nothing the imager can do to add the missing data to a single image, but aesthetically, so long as the results are eye pleasing and the artifacts are limited, this is acceptable.

The brightness and contrast controls can be tweaked to improve the overall appearance, but it is important to have an understanding of how these functions affect a histogram. Brightness functions affect all pixels of an image equally, or linearly. Increasing the brightness moves the total histogram toward the highlight end and away from the shadows. Increase the brightness enough, and clipping occurs in the brighter areas of the image while highlight data becomes lost. The same is true if an image is darkened; movement of the histogram is toward the shadows. Move it enough, and shadow detail becomes clipped. An increase in contrast stretches the histogram of an image and decreasing contrast compresses a histogram. If an image has been corrected to display a full range of tones, any further adjustments through contrast controls will remove data at both the highlight and shadow ends of the histogram. For the most part use the brightness and contrast controls sparingly; a function that controls the histogram in a nonlinear fashion (histogram sliders) is best for tonal corrections.

Image Sharpening

Photo quality is a subjective characteristic that is often determined by the definition or sharpness of an image. Mushy, foggy edges give a photo a dreamlike, unrealistic quality. Earlier in this chapter we mentioned the use in film circles of a retouching pencil to put a hard edge between the light and dark areas of a photo. A similar effect can be obtained digitally with editing software through the function called sharpening.

Sharpening is not the cure-all for blurry images. It is a correction feature for analog images that lose a bit of sharpness when they are converted to a digital format. It is also a correction for the loss of sharpness incurred when images are printed as hardcopy or if resampled. Regardless, most often sharpening is used when an original image is a bit soft in appearance. The process can be quite effective if mildly applied. When a photo is over-sharpened, as can be seen in Figure 8.14, artifacts become apparent, noise is exaggerated, and the image takes on a surrealistic quality. Sharpening works by creating higher contrast between certain neighboring pixels, and creating a hard edge and a snap the image may lack. Experiment with the settings for sharpening using your photo-editing software. Each image will tolerate sharpening effects differently.

To reduce the amount of noise in a sharpened photo it is advantageous to begin the editing process by layering two or more same-quality photos taken as close together in time as possible. Layering images removes hot pixels, boosts contrast, and increases detail slightly. The difficulty when stacking a large number of solar pictures is having image uniformity from one frame to the next because of inconsistent seeing conditions. The daytime sky sometimes is made up of crisp cells of good seeing that are no larger than a small portion of the overall field of view in the camera. One image will contain some areas in sharp focus, while other sections are blurred. The next image can exhibit differing regions that are sharp or blurry. Consequently, layering or stacking is most effective if the series of images

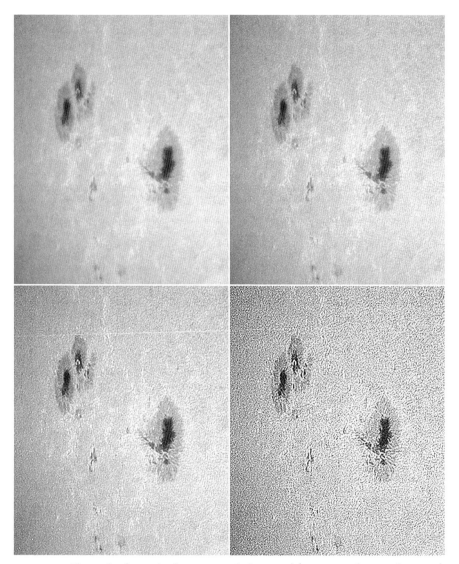

Figure 8.14. The results of using the sharpening tool. The *upper left image* is unsharpened; to its *right* is a mildly sharpened image, and in the *lower left* is a strongly sharpened image. At the *lower right* an over-sharpened image is shown with strong artifacts.

are of similar quality. This is not to say the technique is impossible; several solar observers have demonstrated the positive effects of image stacking, But again, the technique is at the mercy of the seeing conditions.

Saving the File

Once an image has been edited to an eye-pleasing appearance the file should be saved in a loss-less format, such as TIFF. To complete the editing process, information pertinent to the file is embedded as text in the image. This is normally done by adding a border with a wide tab at the foot of the image and, from a prepared template, paste the necessary data within the tab. Data includes the active region number or the position angle if a limb feature, date, time, seeing and transparency conditions, instrumentation used to acquire the image, and an identification of the observer. Lastly, the calibration grayscale and a directional indicator are pasted into the file. The naming convention for the file is left to an imager, but giving the file a name that reflects both the date and universal time the image was obtained is helpful when archiving. For example, 20070815.1715.jpg indicates the image is from 2007 – year, 08 – month, 15 – day, 1715 – UT.

For archiving purposes a folder can be created with a name reflecting the year/month/day of the images stored inside it. This folder is to contain any original file archived in its native format along with the edited file in the loss-less format. From the edited file, copies are generated as JPEGs for Internet uploading or e-mailing to friends. Storing a collection of images strictly on the hard drive of a computer is risky business. To prevent losing data from a crashed hard drive, develop the habit of also creating copies of all your images on an external media source, such as a DVD, CD, or flash drive.

Working with Channels

The wide brightness range of some monochromatic features, for example a filament on the disc verses a limb prominence, requires two exposure settings. However, for the H-alpha imager whose camera produces an RGB file, an interesting technique exists that captures both bright disc detail and faint prominence detail in a single exposure.

Earlier in this chapter we discussed Bayer Masking and how a digital sensor uses red, green, and blue filters in an array to create a color image. Because H-alpha light is pure red, with no transmission of off-band color, you would expect an RGB digital camera to record H-alpha light in only the red channel. Many times this is not the case because of a defect with the camera sensor called leakage, in which light from one channel (in this case the red) spills onto nearby pixels and creates a false color image.

Open a narrow band H-alpha prominence image in photo-editing software displaying all three channels. If the camera suffers from leakage you will see a solar image in each RGB colorspace. The difference between images will be the amount of over or underexposure and visible detail in each. Depending on the exposure, one channel (red) will likely appear overexposed and washed out, except that the limb prominences may show maximum detail. Another channel (green or blue) may contain an impressive amount of disc detail, with no visible limb features. The remaining third channel will likely be underexposed. The photographer will want to find an exposure time that records the prominence and disc detail simultaneously in the two different channels.

Because of sky conditions, the normal routine is to capture a large number of bracketed exposures at different settings, then during editing pick the best of the lot. At least several of these exposures will meet the above criteria. Experience will serve as a future guide to finding the near ideal exposure, and less bracketing will be required. The green channel, because it uses a greater number of pixels, would be the superior of the three for disc detail.

Select an RGB file to work with, and with the editing software split the channels into three separate grayscale files. Save the selected prominence and disc channel files in a loss-less format and close the original RGB image. Now enhance the detail of the prominence channel and the disc channel using the editing functions. Do not rescale the images or resample the pixels at this time. After editing, these files are saved in the loss-less format and are then composited, with the detailed disc image pasted over the overexposed disc image containing the prominences. The result will be dim and bright solar features captured together in a single grayscale picture. At this point, for aesthetic and not observational purposes, many photographers will colorize the image that can now be enlarged or reduced to the needed size.

Colorizing

By now the realization that effective photography of the Sun is accomplished with a grayscale format should have become obvious. We live in a world that is colorful, and black and white imagery harkens back to a romanticized time before the twenty-first century. From a scientific standpoint, a monochrome image is perfectly acceptable in that it provides data that is measurable. When trying to popularize and attract interest in the Sun, though, or to share a solar experience with friends, a little color can go a long way.

Photo-editing software usually contains a function that permits creating an artificially colored image. All programs work a little differently, but the end result is basically the same. Our software preference for colorizing an image is Adobe's Photoshop. The first task with all coloring programs is to open and convert the grayscale file to one that is RGB. In Photoshop there are several routes to achieving a color image. I normally select the Hue/Saturation function, and by experimenting with the sliders find an eye-catching combination. Avoid functions that adjust the brightness of the image, as these will result in a loss of detail by shifting the histogram toward the highlights or shadows. It is important not to over-process an image when colorizing it, or else the shadow areas may become muddy from the added color. Sometimes you can colorize an image using the duotone mode, which allows critical control of the curves (histogram) of each color channel so you can achieve a little extra snap in the highlights. When the desired duotone effect is achieved, the image is converted to RGB and saved in a loss-less format. There is no right or wrong way when adding color to images; it is a subjective process and the end result, if pleasing to the creator, is what is important.

Creating Isophote Images

In Chapter 5, we outlined a technique of exploring sunspot umbra and the surrounding region by constructing what is called an isophote contour map. An isophote is a grayscale or color graphic that depicts levels of similar density in an image. A sunspot umbra is not the uniform black body it appears to the eye, but rather contains clumps and points of intense magnetic strength that stifle convection more so than the umbral and photospheric regions surrounding them. Such fine gradation of tone inside an umbra is difficult to see visually, but it photographs well when seeing conditions are cooperative. By using the power of photo-editing software and digital imagery it is possible to map these regions, making their existence known. Other delicate white light features such as the inner and outer bright rings and weak light bridges can also be enhanced through this process.

A deep umbral image is suitable for core studies, but a normal sunspot exposure produces interesting results as well. In order to differentiate between levels of density in an image, editing software must be capable of isolating pixels having the same or similar value. Not all programs have a function that can do this task. There is a freeware for Macintosh and Windows platforms online at: http://rsb.info.nih. gov/nih-image/ that is superior for creating an isophote contour map. This application, written for a Macintosh computer, is called, *NIH-Image*; an observer with a Windows PC system should download the *ImageJ* software. A large selection of sample data and instruction material is also available from this free site (Figure 8.15).

One of the functions of the NIH-Image program allows the user to create a density slice of a photograph. The "LUT" (Look Up Table) tool is activated to determine the upper and lower density limits of the slice and its location within the 256 levels of gray. With slicing enabled, all pixels within the density range appear highlighted in red, while the background pixels are left unchanged. On screen, start with a medium thickness slice of approximately 15–20 steps within the 256 levels, saving a screen capture of each slice as you adjust the selection through the complete range of levels. These individual screen captures can be imported into animation software to produce an impressive video clip depicting the continuous flow of density inside a sunspot. This flow is reflective of the sunspot's brightness and temperature variations. To create a single isophote map, however, one needs to combine or layer the screen shots into a flattened map. You can use copy and paste techniques with Photoshop to create the graphic map.

One of the most interesting results from this type of work is the demonstrating of the existence of a core region within an umbra. The core is typically only a few

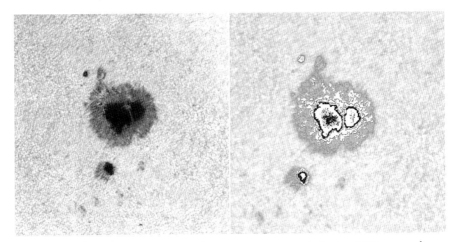

Figure 8.15. The *right image* is an isophote contour map created using NIH-Image software available at http://rsb.info.nih.gov/nih-image/ on a Macintosh G3 computer. In this map the weak light bridge separating the two interior pieces of umbra is clearly visible as well as the "core," or point of greatest density, in the *left* or trailing piece of the umbra.

arc seconds in diameter, with a temperature as much as 500° cooler than that of the surrounding umbra.

Isophote construction is a powerful tool used to illustrate features of a sunspot that one normally overlooks. The availability of the home computer and online freeware make this a method the average solar observer can economically utilize and hardly afford not to. By administering this technique one can easily come away with a greater understanding of the workings of the Sun and its attendant phenomena.

Time-Lapse Videos

Perhaps one of the most exciting activities for an imager is the creation of an animation or video depicting solar morphology. A video clip can speed up motion, in some cases several hundredfold, giving the viewer a clearer picture of the action happening on the Sun. Creation of the video requires an imager to chronologically assemble individual images captured throughout the length of a solar event.

One possibility for subject matter is the birth of a new sunspot group, which documents the emergence of an active region on the Sun's east limb and recording its evolution over the next 2 weeks as it marches across the solar disc. Other possibilities might include imaging the sudden rise and slow decline of a H-alpha or white light flare, and, of course, shooting spectacular prominence activity.

Success in any of the above activities requires three things on the part of the imager: careful planning, commitment, and a bit of luck.

Planning includes having prepared the groundwork for imaging. When capturing images for a video you should not be experimenting with exposure settings, new filters, and so on. You should have had enough experience with your equipment to be able to predict with certainty what the outcome will be from your effort. The first thing you should do when planning a series of images to be turned into a video clip is to develop a timeline of when you intend to capture the pictures. For something like an eruptive prominence, a number of frames (an attempt to capture good seeing) may need to be taken at each point on the timeline. Have a timepiece available, so there is no guesswork as to when it's time for the next image to be taken. As an event progresses check off on the timeline each completed imaging session. An unexpected transient event will have to be quickly planned at the telescope, but a long-duration event can be thought out ahead of time.

Frankly, being in the right place at the right time is important for producing spectacular videos of solar activity. Unlike the professional astronomer, who may dedicate an entire day to a solar observing project, amateurs are usually limited to the leisure time they have available, sometimes only on weekends. Long or short-duration projects require the cooperation of weather patterns. Nothing is more disheartening than to be clouded out just as a flare is peaking and the observer is in the middle of securing images for a video clip of the event. Success in this field is always dependent on a bit of luck.

These projects normally require a commitment of 30 min to several weeks of time. Flare and prominence eruptions are relatively short-lived events in which images are snapped every 15–60 s. Such fast action features usually last from 30 min to several hours at most. However, recording a sunspot group marching across the solar disc can easily be depicted in a video clip made from only a couple of images per day for approximately a two-week period. A smooth appearing video, with minimal "jitters," requires a commitment to obtain images closely and equally spaced for the duration of the event. Holes or gaps resulting from missing images create a jumpy-looking final product.

Depending on the duration of the event and how smooth a video you want to produce, you may need a dozen to several hundred images that are each edited as normal. The most time-consuming portion of the editing phase will be selecting which images from the collection to include in the video. Remember that during each segment on the time line at least several photos must be taken to capture the best seeing. The greater the pixel count of each frame, the larger the final file size will be of your video. So, take into consideration how the video will be put to use. If you intend to e-mail it to your friends, a 2- or 3-megabyte file might be about the maximum file size you want, so plan accordingly. Cropping out unwanted images or resampling each image to a lower resolution may be necessary. Each final image for the video should ideally be of the same scale, resolution, orientation, and quality. Name the files in a numerical sequence so that when they are imported into the animation software, the correct flow is kept. There are many software programs available that perform animation assembly. ImageReady software that comes bundled with Adobe Photoshop is good for video projects. Experimentation on the part of the observer will determine the ideal run time of the video and the amount of fade-in/fade-out needed between frames to achieve the best time compression.

A final suggestion: create a title page at the beginning of the video that is visible for several seconds before the actual action begins playing. The title should inform the viewer of what is seen and how it was obtained. This is a nice touch and embeds the observational data with the clip, making it valuable for archiving.

Reference

1. *Digital Astrophotography: The State of the Art*, D. Ratledge, Springer-Verlag, 2005

Chapter 9

Where do You Go from Here?

Solar astronomy, when explored by an independent amateur astronomer, is a very fulfilling experience, but when it's shared with like-minded individuals, the results become edifying and rewarding for all involved. Perhaps some of the best times I've had observing the Sun were when I was crowded around a friend's telescope with half a dozen other curious observers, with veteran solarphiles like myself talking shop while pointing out a bright flare on the disc or a loop prominence on the limb. Some novice amateurs are not even aware that the Sun can be observed safely, having only heard of the inherent dangers.

If you find yourself much of a lone wolf with this hobby and desire more, begin by seeking out other amateur astronomers through a local club or astronomical society. Many cities have at least one group that meets regularly, conducts monthly observing parties, and welcomes new hobbyists to the ranks. Every group tends to have one or two individuals fostering a similar solar astronomy interest with which camaraderie can begin. You and your new friends can be a resource of knowledge and inspiration for each other and the rest of the club membership. Astronomy Day celebrations will naturally rely on your expertise, because such public events are usually held during the daytime, when the Sun is the only sky object visible. Membership in a local society can last a lifetime.

There are several national and international organizations with sections devoted to solar observing (see Appendix A of this book) and coordinating amateur solar observations. A friendly relationship can be developed through these groups and with other members at the annual conventions they hold. With today's technology, however, much of the communication between the members of a national group occurs via e-mail. Some organizations maintain Internet message boards, on which the membership around the world post questions and answers or contribute correspondence to threads (extended conversations) about specific topics. The primary purpose of some larger groups is to serve as a depository for amateur solar observations, thereby providing access to your observations by the professional astronomer. For instance, the Solar Section of the Association of Lunar and Planetary Observers receives and archives scores of images monthly depicting white light and monochromatic observations from around the world by dozens of solar observers. These observations are made available on-line at the group's website. Other organizations provide similar services. The Solar Division of the American Association of Variable Star Observers tabulates monthly statistical data of the daily sunspot number also from observers worldwide. The British Astronomical Association's Solar Section collects solar observations from astronomers on that side of the Atlantic Ocean. All of these groups coordinate amateur activities for the purpose

J.L. Jenkins, *The Sun and How to Observe It*, DOI 10.1007/978-0-387-09498-4_9,
© Springer Science+Business Media, LLC 2009

of promoting astronomy, educating the public, and providing data to interested researchers, amateur and professional. When you are a part of such a sincere and dedicated group of fellow observers, you are indeed contributing to the knowledge about our Sun, and providing a worthwhile outlet for your own tendencies to pursue science. If you are serious about solar observing, contact an organization that has guidelines that interest you and become an active member.

Many independent amateur solar observers maintain a personal website that focuses on their hobby. These sites perhaps contain information on the equipment an observer has at his or her disposal, detailed instruction on the techniques employed in their observing routine, information about the Sun and their observations, and often a gallery of images illustrating the level of expertise obtained. One observer even has a section devoted to print ads of commercial solar equipment from the past. It's a nice touch, providing a trip down memory lane.

Designing your own site requires little more than web-building software and the rented on-line space from a host-server or your Internet Service Provider (ISP). When considering building a web page, decide what is to be its purpose. Is it to educate those who visit about a particular facet of solar astronomy, or do you wish to focus solely on the details of your hobby? Whatever the goal of your website, establish its purpose at the outset and design it with that in mind. Keep the website fresh and up-to-date. Nothing loses the interest of a visitor faster than to find that the latest update to a site was several years before. It tells the viewer that nothing new is happening with you and your hobby. Fresh and interesting data brings people back time and again.

Appendix A
Resources

Amateur Observing Organizations

Association of Lunar and Planetary Observers Solar Section
www.alpo-astronomy.org
American Association of Variable Star Observers Solar Division
www.aavso.org/observing/programs/solar/
British Astronomical Association Solar Section
www.britastro.org/solar/
Belgian Solar Observer
www.bso.vvs.be/index_en.php

Manufacturers and Suppliers of Solar Equipment

DayStar Filters
www.daystarfilters.com
Coronado Filters
www.coronadofilters.com
Lunt Solar Systems
www.luntsolarsystems.com
Lumicon International
www.lumicon.com
Thousand Oaks Optical
www.thousandoaksoptical.com
Baader Planetarium
www.baader-planetarium.com
Kendrick Astro Instruments
www.kendrickastro.com
Alpine Astronomical
www.alpineastro.com
Seymour Solar
www.seymoursolar.com

Photographic Suppliers

Procyon Systems
www.procyon-systems.com
Starlight Xpress Ltd.
www.starlight-xpress.com.uk
Adorama Inc.
www.adorama.com
Orion Telescopes
www.oriontelescopes.com
The Imaging Source
www.astronomycameras.com
Diffraction Limited
www.cyanogen.com
Santa Barbara Instrument Group
www.sbig.com
Adirondack Astronomy
www.astrovid.com
Lumenera Corporation
www.lumenera.com
Nikon
www.nikonusa.com
Canon
www.usa.canon.com

Appendix B
Glossary of Solar-Related Terms

Absorption lines	The dark lines crossing a spectrum caused by the absorbing of photons as electrons jump to a higher energy level.
Active region	Location in the photosphere in which the formation of sunspots, faculae, etc., arise over time.
Angstrom	Unit of measure for the expression of the wavelength of light equal to .1 nm or one ten-millionth of a millimeter.
Aurora	Glowing gases in Earth's upper atmosphere, excited by solar particles originating in the Sun and carried to Earth by the solar wind.
Bo	A parameter for the calculation of heliographic coordinates representing the nod or changing latitude at the center of the solar disc.
Balmer series	A pattern of spectral lines in the visible spectrum produced by the jumping of electrons from one energy level to another in hydrogen.
Bandpass	Measurement between the lower and upper cut-off frequency of an optical filter, usually measured at the full-width half-maximum point.
Bandwidth	See bandpass.
Bipolar sunspot	Two concentrations of umbral spots or sunspots having a positive and negative magnetic polarity with a minimum separation of 3 heliographic degrees.
Bright point	A tiny dot within an umbra having a brightness greater than the surrounding umbra and nearby umbral dots.
Broadband	Term for a filter transmitting a wide bandwidth (i.e., 100 Å or more).
Butterfly diagram	The graphical representation depicting the latitude of emerging sunspots versus the time progression of a solar cycle.
Calcium-K	The spectral line located at 393.3 nm.
Carrington rotation	The number of rotations of the Sun as seen from Earth since November 9, 1853.
Center wavelength (CWL)	The wavelength found at the midpoint of the full-width half-maximum.
Central meridian (CM)	An imaginary line drawn from the north to the south pole of the Sun.

Chromosphere	The layer of solar atmosphere directly above the photosphere and below the corona.
Chromospheric network	A web-like mesh covering almost the entire Sun and displaying a bright pattern in Ca-K and in H-alpha, a dark one.
Coelostat	Two mirror system reflecting a stationary image of the Sun to a telescope. Also see heliostat.
Convection zone	An inner layer of the Sun in which energy transfer occurs through convection.
Core (Sun)	The central region of the solar interior powered by the hydrogen to helium fusion process.
Core (sunspot)	The area of a sunspot umbra having the greatest magnetic strength.
Corona	The outer atmosphere of the Sun, beyond the chromosphere.
Coronagraph	An instrument designed to permit viewing of the Sun's corona.
Coronal mass ejection (CME)	A large eruption of particles from the Sun.
Differential rotation	The lack of singularity in a rotation period due to the liquid-like nature of the body.
Diffraction grating	A finely grooved substrate whose purpose is the dispersion of light.
Disparition brusque	The sudden disappearance of a filament (prominence).
Dobsonian solar telescope	A unique Newtonian-style telescope designed for white light solar observing, the creation of sidewalk astronomer, John Dobson.
Doppler shift	The stretching or compressing of spectral lines due to the approaching or receding of an object.
Double stack	A method of narrowing the bandwidth of an etalon by the addition of a second etalon.
Ellerman bomb	A small bright feature visible in the wings of H-alpha notably around a sunspot. Circular with a diameter less than 5 s, they have a lifetime of a few minutes to several hours in rare instances.
Emission lines	The bright lines crossing a spectrum caused by the emitting of photons as electrons jump to a lower energy level.
End-loading	A monochromatic narrow band filter that attaches at the exit of a telescope.
Energy rejection filter (ERF)	A pre-filter that is placed over the opening of a telescope for the purpose of absorbing or reflecting UV/IR light and reducing the heat load on the interference filter.
Ephemeres	Tables that are published yearly listing the daily orientation of the Sun for the factors P, Bo, and Lo.
Eruptive prominence	An active prominence being ejected from the Sun.
Etalon	An optical filter that operates by the multiple-beam interference of light, reflected and transmitted by a pair of parallel flat reflecting plates.

Facula	(pl. faculae) A luminous, cloud-like patch or venous streak of material surrounding or near a sunspot.
Fibril	A tiny, dark-appearing structure that follows a magnetic field line; sometimes attached to a prominence seen on the solar disc. Analogous to a mottle that is longer than a few seconds of arc.
Field angle	The angle of outside light rays entering a telescope. One example is illustrated by the angular size of the Sun as it appears in the sky.
Filament	In monochromatic light a prominence viewed on the disc of the Sun. In white light a structure that radiates about an umbra like fine, dark threads.
Filar micrometer	A tool for measuring angular displacement through an optical instrument.
Filigree	Tiny bright flux tubes that pop through the solar surface with a diameter of about 150 km.
Flash phase	The period of rapid brightening experienced in a solar flare.
Flux tube	A strand or kink of magnetic field suspended in the convection zone.
Fraunhofer line	An atomic line visible in a spectrum.
Front-loading	A monochromatic narrow band filter that attaches at the entrance of a telescope.
Full-width half-maximum (FWHM)	The measured width of the bandpass, in nanometers or angstroms, at one-half of the maximum transmission.
G-band	The location of several spectral lines that go into emission during a flare at about 430 nm.
Granulation	The textured pattern found over the entire photospheric surface of the Sun. See *granule.*
Granule	The top of a rising column of gas, originating deep within the convection zone of the Sun.
H-alpha	The spectral line located at 656.3 nm.
Heliographic coordinates	The system of latitude and longitude on the solar disc.
Helioseismology	The study of low frequency sound waves originating in the Sun.
Heliostat	A single- or multiple-mirror system reflecting an image of the Sun to an optical instrument.
Helmholtz contraction	The process of turning gravity's energy to heat as induced by density and pressure.
Herschel wedge	A narrow prism used for safely observing the white light Sun when combined with suitable supplementary filters.
Hossfield pyramid	The name given to a pyramid-shaped projection box for white light solar observing; devised by Casper Hossfield of the AAVSO Solar Division.
Hydrogen	The most plentiful element in the Sun and universe, made of one proton and one electron.

Infrared (IR)	Electromagnetic radiation with a wavelength between approximately 700 nm and 1 mm.
Ion	An atom having one or more electrons missing.
Inner bright ring	A brightening within the penumbra located between the umbra and penumbra at a rough region where penumbral filaments have the appearance of extensions of the umbra.
Instrument angle	The angle of light rays converging to a focus in a telescope.
Interference filter	An optical appliance with several layers of evaporated coatings on a substrate, whose spectral transmission characteristics are the result of the interference of light rather than absorption.
Intergranular wall	The intergranular wall is what defines the shape of a granule.
Irregular penumbra	A sunspot penumbra that has been mutated by complex magnetic fields.
Isophote contour map	A graphically created image that interprets the many levels of density within a photo.
Kelvin	A unit of temperature in which zero Kelvin is based on $-273.15°$C or absolute zero.
Kirchhoff's laws of spectroscopy	Laws that state that (1) A hot, dense, glowing body produces a continuous spectrum lacking spectral lines. (2) View a continuous spectrum through a cooler, transparent gas, and dark lines called absorption lines appear. (3) Hot, transparent gas before a cooler background emits the bright spectral lines we call emission lines.
Lo	A parameter for the calculation of heliographic coordinates representing the longitude of the Sun's central meridian. *Lo* is $0°$ at the beginning of each new solar rotation.
Light bridge	Any material brighter than an umbra that also divides the umbra, often times dividing even a penumbra.
Limb darkening	The decrease in intensity of the Sun as one approaches the solar limb.
Lyot filter	A monochromatic filter that produces a narrow band transmittance via the principle of birefringence.
Magnetic cycle	The return of the magnetic polarity of sunspots within a given hemisphere to the polarity experienced prior to a new solar cycle. A period of approximately 22 years.
Magnetosphere	The vicinity around Earth dominated by its magnetic field.
Magnitude	A measure of brightness of a celestial body.
McIntosh classification	A three-digit white light sunspot classification scheme devised for flare prediction.
Mean daily frequency (MDF)	An index of solar activity determined by the number of active regions visible as the average for a monthly period.

Menzel-Evans classification	A prominence classification system based on whether a prominence is ascending or descending in the chromosphere, its relationship to any nearby sunspot, and the general appearance of the prominence.
Monochromatic	One color, as when referring to light from the H-alpha or Calcium K-line.
Moreton wave	Shock wave visible in the chromosphere radiating from a large flare.
Morphology	The study of the changing appearance of the Sun.
Mottle	A spicule seen against the solar disc. Also see *fibril*.
Nanometer	The nanometer (nm) is a unit measurement of wavelengths of electromagnetic radiation (light). One nanometer is equal to one billionth of a meter (1×10^{-9} m).
Narrow band	Term for a filter transmitting a narrow bandwidth (i.e., less than 100 Å).
Neutral line	The area where an active region's magnetic field reverses polarity.
Normal incidence	Light rays that have a normal or parallel path.
Objective filter	White light solar filter that mounts at the entrance to a telescope.
Occulting cone	Polished metal cone-shaped devise to block the light from the solar disc in a prominence telescope.
Off-axis	In solar observing the placement of a sub-diameter objective filter at the side of the optical axis of a telescope to avoid the internal optical components (secondary mirror, mounting hardware, etc.) from being within the incoming light path.
Outer bright ring	The brightening and aligning of the granules encircling the outer edge of a sunspot, beyond the penumbra.
Oven	An electrically controlled heating device to regulate the operating temperature of a narrow band monochromatic filter.
P	Parameter for the calculation of heliographic coordinates representing the displacement of the north rotational axis of the Sun relative to the rotational axis of Earth.
Peak transmission	The maximum percentage of transmission found within the bandwidth.
Penumbra	The lighter, grayish outer region surrounding the umbra in a sunspot.
Penumbral filaments	Structures of fine dark threads that radiate about a sunspot umbra.
Penumbral grains	Bright regions located between penumbral filaments.
Photon	A particle that supports electromagnetic radiation.
Photosphere	The lowest layer of the Sun's atmosphere. The region contains sunspots, granulation, and faculae.

Plage	A lower chromosphere feature that frequently surrounds a sunspot as a bright, cloud-like mass. It marks the location of the magnetic field associated with the sunspot.
Plasma	A brew of ions and electrons that react energetically with magnetic fields.
Polar crown	The appearance of several filaments linking together in an east-west direction to form one long strand of filamentary material in a high solar latitude.
Polar faculae	Small bits of faculae forming in high latitude regions outside the sunspot zones.
Pore	A tiny structure, with a diameter from 1 to 5 arc seconds, darker than a granule but brighter than the umbra of a well-developed sunspot.
Position angle (PA)	The angular offset in degrees around the limb of the Sun defining the position of a prominence or other feature. The cardinal directions are N=0°, E=90°, S=180, and W=270°.
Prominence	A cloud of gas suspended above the surface of the Sun.
Proton-proton cycle	The process experienced by a star equal to or less than the mass of the Sun by which it converts hydrogen to helium.
Quiescent prominence	A quiet behaving prominence that changes its appearance moderately with time.
Radiative zone	An inner layer of the Sun in which energy transfer occurs through radiative properties.
Relative sunspot number	An index of solar activity determined by the number of sunspot groups and individual sunspots visible on the face of the Sun.
Rudimentary penumbra	The beginning phase of a penumbra, forming from the intergranular material, surrounding a newly developed umbra.
Seeing conditions	The quality of the atmosphere between an observer and what is being viewed.
Sidereal period	The time required for the Sun to complete one rotation, as seen from a fixed point in space.
Solar continuum	Basically a view of the Sun encompassing a wide bandwidth of visible light. It would be analogous to a white light view.
Solar cycle	The approximate 11-year rise and fall in solar activity.
Solar flare	The swift release of energy that has accumulated within the magnetic field of an active region.
Solar maximum	The peak of solar activity during the 11-year solar cycle.
Solar minimum	The time when the Sun experiences little activity during the 11-year solar cycle.
Solar nebula	A vast cloud of gas and dust from which it is believed the Sun and Solar System originated.

Solar projection	A technique for viewing the white light Sun by projecting an enlarged image of the solar disc onto a white screen some distance from the eyepiece of a telescope.
Solar wind	The constant stream of particles flowing from the Sun into space.
Space weather	The status of the space environment near Earth, as it has been affected by the release of energy and particles from the Sun.
Spectrohelioscope	An instrument that synthesizes a monochromatic view of the Sun.
Spectroscope	An instrument for the dispersion of light.
Spectrum	The result when electromagnetic energy is dispersed into its constituent rays by order of wavelength, for example, a rainbow.
Spicule	A fine structure resembling a tiny gas jet at the solar limb. On the disc in H-alpha it appears dark and then is known as a mottle or fibril.
Star	A giant sphere-shaped ball of gas that, through nuclear reactions, releases energy in its core.
Stonyhurst Disc	A template or grid that shows the lines of heliographic latitude and longitude relative to a given value of Bo.
Sunspot	A dark region on the photosphere that results from convection being stifled by magnetic fields within the region.
Sunspot drift	The passage of a sunspot or sunspot group from east to west as seen in a stationary telescope, caused by Earth's rotation.
Sunspot group	A clump of sunspots.
Sunspot zone	A region of approximately $35°$ on both sides of the solar equator in which sunspots form.
Supergranulation	The large-scale pattern in the photosphere of organized cellular structure. Each cell contains hundreds of individual granules and has a diameter around 30,000 km. The chromospheric network overlays the supergranulation pattern.
Synodic period	The apparent rotation of the Sun, as seen from the orbiting Earth.
Tachocline	The region located between the radiative and convection zones of the solar interior.
Telecentric lens	A supplementary lens system intended to create normal incidence light rays from the converging light rays of a telescope.
Transparency	A characteristic of sky conditions that describes the opacity of the atmosphere as influenced by water vapor, dust, smoke, and other atmospheric particles.
Ultraviolet (UV)	Electromagnetic radiation with a wavelength between approximately 400 nm and 10 nm.
Umbra	The dark, cooler region of a sunspot.

Umbra dot	A small dot within an umbra having a brightness between that of the umbra and nearby bright points.
Umbra spot	A pore that has become larger and darker than other pores, as dark as the typical sunspot umbra.
Unipolar sunspot	A single concentration of umbral spots or sunspots within a 3-degree or less area.
Universal Time	Analogous to Greenwich Mean Time, a 24-hour timescale.
Visible light	Electromagnetic radiation with a wavelength between approximately 400 nm and 700 nm.
Wave front error	A distortion or aberration to the incoming rays of light caused by factors including poor atmospheric seeing or poorly made optics.
Wavelength	The distance separating two consecutive wave peaks in a beam of electromagnetic radiation (light).
White light	The combined result of all wavelengths of visible light.
White light flare (WLF)	A solar flare of such intensity that its light outshines the solar continuum and becomes visible without a monochromatic filter.
Wien's law	The principle that the wavelength of the dominant color of a blackbody (star), multiplied by its temperature must equal a specific numerical factor.
Wilson effect	The apparent concavity of a symmetrically shaped sunspot as it is near the solar limb.
X-ray	Electromagnetic radiation with a wavelength between 10 nm and 0.01 nm.

Appendix C Daily Solar Ephemeris, July 2008–January 2012

CALENDAR DATE	ROTATION NUMBER	HELIOGRAPHIC			DIAMETER	RA	DEC
		Lo	Bo	Po	(arcmin)	(HH:MM)	(deg)
7/ 1/2008	2071	79.17	2.90	−2.54	31.463	06:41.9	23.10
7/ 2/2008	2071	65.99	3.01	−2.09	31.462	06:46.1	23.03
7/ 3/2008	2071	52.75	3.12	−1.63	31.462	06:50.2	22.95
7/ 4/2008	2071	39.51	3.23	−1.18	31.462	06:54.3	22.86
7/ 5/2008	2071	26.27	3.34	−0.73	31.462	06:58.4	22.77
7/ 6/2008	2071	13.03	3.44	−0.27	31.462	07: 2.5	22.67
7/ 7/2008	2071	359.79	3.55	0.18	31.462	07: 6.7	22.57
7/ 8/2008	2072	346.56	3.65	0.63	31.463	07:10.7	22.46
7/ 9/2008	2072	333.32	3.76	1.08	31.463	07:14.8	22.34
7/10/2008	2072	320.08	3.86	1.53	31.464	07:18.9	22.22
7/11/2008	2072	306.84	3.96	1.98	31.465	07:23.0	22.09
7/12/2008	2072	293.60	4.06	2.43	31.466	07:27.1	21.95
7/13/2008	2072	280.37	4.16	2.88	31.467	07:31.1	21.81
7/14/2008	2072	267.13	4.26	3.32	31.469	07:35.2	21.66
7/15/2008	2072	253.89	4.35	3.76	31.470	07:39.2	21.50
7/16/2008	2072	240.66	4.45	4.21	31.472	07:43.3	21.34
7/17/2008	2072	227.42	4.54	4.65	31.474	07:47.3	21.17
7/18/2008	2072	214.18	4.64	5.08	31.475	07:51.3	21.00
7/19/2008	2072	200.95	4.73	5.52	31.478	07:55.3	20.82
7/20/2008	2072	187.71	4.82	5.95	31.480	07:59.3	20.64
7/21/2008	2072	174.48	4.91	6.38	31.482	08: 3.3	20.44
7/22/2008	2072	161.31	5.00	6.81	31.485	08: 7.3	20.25
7/23/2008	2072	148.08	5.09	7.24	31.487	08:11.3	20.04
7/24/2008	2072	134.84	5.17	7.66	31.490	08:15.2	19.84
7/25/2008	2072	121.61	5.25	8.08	31.493	08:19.2	19.62
7/26/2008	2072	108.38	5.34	8.50	31.496	08:23.1	19.40
7/27/2008	2072	95.15	5.42	8.91	31.499	08:27.1	19.18
7/28/2008	2072	81.91	5.50	9.33	31.503	08:31.0	18.95
7/29/2008	2072	68.68	5.58	9.73	31.506	08:34.9	18.72
7/30/2008	2072	55.45	5.65	10.14	31.510	08:38.8	18.48
7/31/2008	2072	42.22	5.73	10.54	31.513	08:42.7	18.23
8/ 1/2008	2072	28.99	5.80	10.94	31.517	08:46.6	17.98
8/ 2/2008	2072	15.77	5.87	11.33	31.521	08:50.5	17.73
8/ 3/2008	2072	2.54	5.94	11.72	31.526	08:54.3	17.47
8/ 4/2008	2073	349.31	6.01	12.11	31.530	08:58.2	17.20
8/ 5/2008	2073	336.08	6.08	12.49	31.534	09: 2.0	16.93

(continued)

CALENDAR DATE	ROTATION NUMBER	HELIOGRAPHIC			DIAMETER	RA	DEC
		Lo	Bo	Po	(arcmin)	(HH:MM)	(deg)
8/ 6/2008	2073	322.86	6.14	12.87	31.539	09: 5.9	16.66
8/ 7/2008	2073	309.63	6.21	13.25	31.543	09: 9.7	16.38
8/ 8/2008	2073	296.40	6.27	13.62	31.548	09:13.5	16.10
8/ 9/2008	2073	283.18	6.33	13.99	31.553	09:17.3	15.81
8/10/2008	2073	269.95	6.39	14.35	31.558	09:21.1	15.52
8/11/2008	2073	256.79	6.44	14.71	31.564	09:24.9	15.23
8/12/2008	2073	243.57	6.50	15.06	31.569	09:28.7	14.93
8/13/2008	2073	230.35	6.55	15.41	31.574	09:32.4	14.63
8/14/2008	2073	217.12	6.60	15.76	31.580	09:36.2	14.32
8/15/2008	2073	203.90	6.65	16.10	31.585	09:39.9	14.01

CALENDAR DATE	ROTATION NUMBER	HELIOGRAPHIC			DIAMETER	RA	DEC
		Lo	Bo	Po	(arcmin)	(HH:MM)	(deg)
8/16/2008	2073	190.68	6.70	16.44	31.591	09:43.7	13.69
8/17/2008	2073	177.46	6.75	16.77	31.597	09:47.4	13.38
8/18/2008	2073	164.24	6.79	17.09	31.603	09:51.1	13.05
8/19/2008	2073	151.02	6.83	17.42	31.609	09:54.9	12.73
8/20/2008	2073	137.80	6.87	17.73	31.616	09:58.6	12.40
8/21/2008	2073	124.58	6.91	18.05	31.622	10: 2.3	12.07
8/22/2008	2073	111.37	6.95	18.35	31.628	10: 6.0	11.73
8/23/2008	2073	98.15	6.98	18.66	31.635	10: 9.6	11.40
8/24/2008	2073	84.93	7.01	18.95	31.642	10:13.3	11.06
8/25/2008	2073	71.72	7.04	19.25	31.648	10:17.0	10.71
8/26/2008	2073	58.50	7.07	19.53	31.655	10:20.7	10.36
8/27/2008	2073	45.29	7.10	19.81	31.662	10:24.3	10.01
8/28/2008	2073	32.07	7.12	20.09	31.669	10:28.0	9.66
8/29/2008	2073	18.86	7.14	20.36	31.677	10:31.6	9.31
8/30/2008	2073	5.64	7.16	20.63	31.684	10:35.3	8.95
8/31/2008	2074	352.49	7.18	20.89	31.691	10:38.9	8.59
9/ 1/2008	2074	339.28	7.20	21.14	31.699	10:42.5	8.23
9/ 2/2008	2074	326.07	7.21	21.39	31.706	10:46.1	7.87
9/ 3/2008	2074	312.86	7.22	21.63	31.714	10:49.8	7.50
9/ 4/2008	2074	299.65	7.23	21.87	31.722	10:53.4	7.13
9/ 5/2008	2074	286.44	7.24	22.10	31.729	10:57.0	6.76
9/ 6/2008	2074	273.23	7.24	22.33	31.737	11:0 .6	6.39
9/ 7/2008	2074	260.02	7.25	22.55	31.745	11: 4.2	6.01
9/ 8/2008	2074	246.81	7.25	22.77	31.753	11: 7.8	5.64
9/ 9/2008	2074	233.60	7.25	22.98	31.761	11:11.4	5.26
9/10/2008	2074	220.40	7.25	23.18	31.770	11:15.0	4.88
9/11/2008	2074	207.19	7.24	23.38	31.778	11:18.6	4.50
9/12/2008	2074	193.98	7.23	23.57	31.786	11:22.2	4.12
9/13/2008	2074	180.78	7.23	23.75	31.795	11:25.8	3.74

(continued)

CALENDAR DATE	ROTATION NUMBER	HELIOGRAPHIC			DIAMETER	RA	DEC
		Lo	Bo	Po	(arcmin)	(HH:MM)	(deg)
9/14/2008	2074	167.57	7.21	23.93	31.803	11:29.4	3.36
9/15/2008	2074	154.37	7.20	24.10	31.811	11:33.0	2.97
9/16/2008	2074	141.16	7.19	24.27	31.820	11:36.5	2.58
9/17/2008	2074	127.96	7.17	24.43	31.829	11:40.1	2.20
9/18/2008	2074	114.75	7.15	24.59	31.837	11:43.7	1.81
9/19/2008	2074	101.55	7.13	24.73	31.846	11:47.3	1.42
9/20/2008	2074	88.41	7.10	24.87	31.855	11:50.9	1.03
9/21/2008	2074	75.21	7.08	25.01	31.864	11:54.5	0.65
9/22/2008	2074	62.01	7.05	25.14	31.873	11:58.1	0.26
9/23/2008	2074	48.81	7.02	25.26	31.881	12: 1.7	−0.13
9/24/2008	2074	35.61	6.99	25.37	31.890	12: 5.3	−0.52
9/25/2008	2074	22.41	6.95	25.48	31.899	12: 8.9	−0.91
9/26/2008	2074	9.21	6.92	25.58	31.908	12:12.5	−1.30
9/27/2008	2075	356.01	6.88	25.68	31.917	12:16.1	−1.69
9/28/2008	2075	342.81	6.84	25.77	31.927	12:19.7	−2.08
9/29/2008	2075	329.61	6.80	25.85	31.936	12:23.3	−2.47
9/30/2008	2075	316.41	6.75	25.92	31.945	12:26.9	−2.86

CALENDAR DATE	ROTATION NUMBER	HELIOGRAPHIC			DIAMETER	RA	DEC
		Lo	Bo	Po	(arcmin)	(HH:MM)	(deg)
10/ 1/2008	2075	303.21	6.71	25.99	31.954	12:30.5	−3.25
10/ 2/2008	2075	290.01	6.66	26.05	31.963	12:34.2	−3.64
10/ 3/2008	2075	276.82	6.61	26.11	31.972	12:37.8	−4.02
10/ 4/2008	2075	263.62	6.55	26.15	31.981	12:41.4	−4.41
10/ 5/2008	2075	250.42	6.50	26.19	31.991	12:45.1	−4.80
10/ 6/2008	2075	237.23	6.44	26.23	32.000	12:48.7	−5.18
10/ 7/2008	2075	224.03	6.38	26.25	32.009	12:52.4	−5.56
10/ 8/2008	2075	210.84	6.32	26.27	32.018	12:56.0	−5.95
10/ 9/2008	2075	197.64	6.26	26.28	32.028	12:59.7	−6.33
10/10/2008	2075	184.51	6.20	26.28	32.037	13: 3.4	−6.71
10/11/2008	2075	171.31	6.13	26.28	32.046	13: 7.1	−7.08
10/12/2008	2075	158.12	6.06	26.27	32.055	13:10.8	−7.46
10/13/2008	2075	144.93	5.99	26.25	32.064	13:14.5	−7.83
10/14/2008	2075	131.73	5.92	26.22	32.073	13:18.2	−8.21
10/15/2008	2075	118.54	5.85	26.19	32.083	13:21.9	−8.58
10/16/2008	2075	105.35	5.77	26.15	32.092	13:25.6	−8.95
10/17/2008	2075	92.15	5.70	26.10	32.101	13:29.4	−9.31
10/18/2008	2075	78.96	5.62	26.04	32.110	13:33.1	−9.68
10/19/2008	2075	65.77	5.54	25.98	32.119	13:36.9	−10.04
10/20/2008	2075	52.58	5.46	25.91	32.128	13:40.6	−10.40
10/21/2008	2075	39.39	5.37	25.83	32.137	13:44.4	−10.76

(continued)

CALENDAR DATE	ROTATION NUMBER	HELIOGRAPHIC			DIAMETER	RA	DEC
		Lo	Bo	Po	(arcmin)	(HH:MM)	(deg)
10/22/2008	2075	26.19	5.29	25.74	32.146	13:48.2	−11.11
10/23/2008	2075	13.00	5.20	25.64	32.155	13:52.0	−11.46
10/24/2008	2075	359.81	5.11	25.54	32.164	13:55.8	−11.81
10/25/2008	2076	346.62	5.02	25.43	32.172	13:59.6	−12.16
10/26/2008	2076	333.43	4.93	25.31	32.181	14: 3.5	−12.50
10/27/2008	2076	320.24	4.84	25.19	32.190	14: 7.3	−12.84
10/28/2008	2076	307.05	4.74	25.05	32.199	14:11.2	−13.18
10/29/2008	2076	293.86	4.65	24.91	32.207	14:15.1	−13.51
10/30/2008	2076	280.67	4.55	24.76	32.216	14:18.9	−13.84
10/31/2008	2076	267.55	4.45	24.60	32.224	14:22.8	−14.17
11/ 1/2008	2076	254.36	4.35	24.44	32.233	14:26.8	−14.49
11/ 2/2008	2076	241.17	4.25	24.27	32.241	14:30.7	−14.80
11/ 3/2008	2076	227.98	4.14	24.08	32.249	14:34.6	−15.12
11/ 4/2008	2076	214.79	4.04	23.89	32.258	14:38.6	−15.43
11/ 5/2008	2076	201.60	3.93	23.70	32.266	14:42.5	−15.73
11/ 6/2008	2076	188.42	3.83	23.49	32.274	14:46.5	−16.04
11/ 7/2008	2076	175.23	3.72	23.28	32.282	14:50.5	−16.33
11/ 8/2008	2076	162.04	3.61	23.06	32.290	14:54.5	−16.62
11/ 9/2008	2076	148.85	3.50	22.83	32.297	14:58.5	−16.91
11/10/2008	2076	135.67	3.39	22.60	32.305	15: 2.6	−17.19
11/11/2008	2076	122.48	3.27	22.35	32.313	15: 6.6	−17.47
11/12/2008	2076	109.29	3.16	22.10	32.320	15:10.7	−17.74
11/13/2008	2076	96.11	3.04	21.84	32.328	15:14.8	−18.01
11/14/2008	2076	82.92	2.93	21.58	32.335	15:18.9	−18.27
11/15/2008	2076	69.74	2.81	21.30	32.342	15:23.0	−18.53

CALENDAR DATE	ROTATION NUMBER	HELIOGRAPHIC			DIAMETER	RA	DEC
		Lo	Bo	Po	(arcmin)	(HH:MM)	(deg)
11/16/2008	2076	56.55	2.69	21.02	32.349	15:27.1	−18.78
11/17/2008	2076	43.37	2.57	20.73	32.356	15:31.2	−19.03
11/18/2008	2076	30.18	2.45	20.44	32.363	15:35.4	−19.27
11/19/2008	2076	17.00	2.33	20.13	32.370	15:39.5	−19.50
11/20/2008	2076	3.87	2.21	19.82	32.377	15:43.7	−19.73
11/21/2008	2077	350.69	2.09	19.50	32.383	15:47.9	−19.95
11/22/2008	2077	337.51	1.97	19.18	32.390	15:52.1	−20.17
11/23/2008	2077	324.32	1.85	18.85	32.396	15:56.3	−20.38
11/24/2008	2077	311.14	1.72	18.51	32.402	16:0 .5	−20.59
11/25/2008	2077	297.95	1.60	18.16	32.408	16: 4.8	−20.78
11/26/2008	2077	284.77	1.47	17.81	32.414	16: 9.0	−20.98
11/27/2008	2077	271.59	1.35	17.45	32.420	16:13.3	−21.16
11/28/2008	2077	258.40	1.22	17.09	32.426	16:17.6	−21.34
11/29/2008	2077	245.22	1.10	16.72	32.431	16:21.9	−21.51

(continued)

CALENDAR DATE	ROTATION NUMBER	HELIOGRAPHIC			DIAMETER	RA	DEC
		Lo	Bo	Po	(arcmin)	(HH:MM)	(deg)
11/30/2008	2077	232.04	0.97	16.34	32.437	16:26.2	−21.67
12/ 1/2008	2077	218.86	0.84	15.96	32.442	16:30.5	−21.83
12/ 2/2008	2077	205.68	0.72	15.57	32.447	16:34.8	−21.98
12/ 3/2008	2077	192.49	0.59	15.17	32.452	16:39.1	−22.13
12/ 4/2008	2077	179.31	0.46	14.77	32.457	16:43.5	−22.26
12/ 5/2008	2077	166.13	0.33	14.36	32.462	16:47.8	−22.39
12/ 6/2008	2077	152.95	0.20	13.95	32.467	16:52.2	−22.51
12/ 7/2008	2077	139.77	0.08	13.54	32.471	16:56.5	−22.63
12/ 8/2008	2077	126.59	−0.05	13.11	32.475	17:0.9	−22.74
12/ 9/2008	2077	113.41	−0.18	12.69	32.479	17: 5.3	−22.84
12/10/2008	2077	100.29	−0.31	12.26	32.483	17: 9.7	−22.93
12/11/2008	2077	87.11	−0.44	11.82	32.487	17:14.1	−23.01
12/12/2008	2077	73.93	−0.56	11.38	32.491	17:18.5	−23.09
12/13/2008	2077	60.75	−0.69	10.94	32.495	17:22.9	−23.16
12/14/2008	2077	47.58	−0.82	10.49	32.498	17:27.3	−23.22
12/15/2008	2077	34.40	−0.95	10.04	32.501	17:31.7	−23.28
12/16/2008	2077	21.22	−1.07	9.58	32.504	17:36.2	−23.32
12/17/2008	2077	8.04	−1.20	9.12	32.507	17:40.6	−23.36
12/18/2008	2078	354.86	−1.33	8.66	32.510	17:45.0	−23.39
12/19/2008	2078	341.68	−1.45	8.20	32.512	17:49.5	−23.42
12/20/2008	2078	328.51	−1.58	7.73	32.515	17:53.9	−23.43
12/21/2008	2078	315.33	−1.70	7.26	32.517	17:58.3	−23.44
12/22/2008	2078	302.15	−1.83	6.79	32.519	18: 2.8	−23.44
12/23/2008	2078	288.98	−1.95	6.31	32.521	18: 7.2	−23.43
12/24/2008	2078	275.80	−2.08	5.84	32.523	18:11.6	−23.42
12/25/2008	2078	262.62	−2.20	5.36	32.525	18:16.1	−23.39
12/26/2008	2078	249.45	−2.32	4.88	32.526	18:20.5	−23.36
12/27/2008	2078	236.27	−2.44	4.40	32.527	18:25.0	−23.32
12/28/2008	2078	223.10	−2.56	3.91	32.528	18:29.4	−23.28
12/29/2008	2078	209.92	−2.68	3.43	32.529	18:33.8	−23.22
12/30/2008	2078	196.81	−2.80	2.95	32.530	18:38.2	−23.16
12/31/2008	2078	183.63	−2.92	2.46	32.531	18:42.7	−23.09

CALENDAR DATE	ROTATION NUMBER	HELIOGRAPHIC			DIAMETER	RA	DEC
		Lo	Bo	Po	(arcmin)	(HH:MM)	(deg)
1/ 1/2009	2078	170.46	−3.04	1.98	32.531	18:47.1	−23.01
1/ 2/2009	2078	157.29	−3.15	1.49	32.531	18:51.5	−22.92
1/ 3/2009	2078	144.11	−3.27	1.01	32.531	18:55.9	−22.83
1/ 4/2009	2078	130.94	−3.38	0.52	32.531	18:60.3	−22.73
1/ 5/2009	2078	117.76	−3.50	0.04	32.531	19: 4.7	−22.62
1/ 6/2009	2078	104.59	−3.61	−0.44	32.531	19: 9.1	−22.51

(continued)

CALENDAR DATE	ROTATION NUMBER	HELIOGRAPHIC			DIAMETER	RA	DEC
		Lo	Bo	Po	(arcmin)	(HH:MM)	(deg)
1/ 7/2009	2078	91.42	−3.72	−0.93	32.530	19:13.4	−22.38
1/ 8/2009	2078	78.25	−3.83	−1.41	32.529	19:17.8	−22.25
1/ 9/2009	2078	65.07	−3.94	−1.89	32.528	19:22.2	−22.11
1/10/2009	2078	51.90	−4.05	−2.37	32.527	19:26.5	−21.97
1/11/2009	2078	38.73	−4.15	−2.84	32.526	19:30.8	−21.82
1/12/2009	2078	25.56	−4.26	−3.32	32.525	19:35.2	−21.66
1/13/2009	2078	12.39	−4.36	−3.79	32.523	19:39.5	−21.49
1/14/2009	2079	359.21	−4.46	−4.26	32.521	19:43.8	−21.32
1/15/2009	2079	346.04	−4.56	−4.73	32.519	19:48.1	−21.14
1/16/2009	2079	332.87	−4.66	−5.20	32.517	19:52.4	−20.95
1/17/2009	2079	319.70	−4.76	−5.66	32.515	19:56.7	−20.76
1/18/2009	2079	306.53	−4.86	−6.13	32.513	20:0 .9	−20.56
1/19/2009	2079	293.42	−4.95	−6.58	32.510	20: 5.2	−20.35
1/20/2009	2079	280.25	−5.04	−7.04	32.507	20: 9.4	−20.14
1/21/2009	2079	267.08	−5.14	−7.49	32.504	20:13.7	−19.92
1/22/2009	2079	253.91	−5.23	−7.94	32.501	20:17.9	−19.70
1/23/2009	2079	240.74	−5.31	−8.39	32.498	20:22.1	−19.46
1/24/2009	2079	227.57	−5.40	−8.83	32.495	20:26.3	−19.23
1/25/2009	2079	214.40	−5.49	−9.27	32.491	20:30.4	−18.98
1/26/2009	2079	201.23	−5.57	−9.70	32.487	20:34.6	−18.73
1/27/2009	2079	188.06	−5.65	−10.13	32.484	20:38.8	−18.48
1/28/2009	2079	174.89	−5.73	−10.56	32.480	20:42.9	−18.22
1/29/2009	2079	161.72	−5.81	−10.98	32.475	20:47.0	−17.95
1/30/2009	2079	148.55	−5.88	−11.40	32.471	20:51.1	−17.68
1/31/2009	2079	135.38	−5.96	−11.81	32.467	20:55.2	−17.41
2/ 1/2009	2079	122.21	−6.03	−12.22	32.462	20:59.3	−17.13
2/ 2/2009	2079	109.04	−6.10	−12.63	32.457	21: 3.4	−16.84
2/ 3/2009	2079	95.87	−6.17	−13.03	32.453	21: 7.4	−16.55
2/ 4/2009	2079	82.70	−6.24	−13.42	32.447	21:11.5	−16.25
2/ 5/2009	2079	69.53	−6.30	−13.81	32.442	21:15.5	−15.95
2/ 6/2009	2079	56.36	−6.36	−14.20	32.437	21:19.5	−15.64
2/ 7/2009	2079	43.19	−6.42	−14.58	32.432	21:23.5	−15.33
2/ 8/2009	2079	30.08	−6.48	−14.96	32.426	21:27.5	−15.02
2/ 9/2009	2079	16.91	−6.54	−15.33	32.420	21:31.5	−14.70
2/10/2009	2079	3.74	−6.59	−15.69	32.415	21:35.5	−14.38
2/11/2009	2080	350.57	−6.65	−16.05	32.409	21:39.4	−14.05
2/12/2009	2080	337.40	−6.70	−16.41	32.402	21:43.4	−13.72
2/13/2009	2080	324.23	−6.74	−16.75	32.396	21:47.3	−13.39
2/14/2009	2080	311.06	−6.79	−17.10	32.390	21:51.2	−13.05
2/15/2009	2080	297.89	−6.83	−17.44	32.384	21:55.1	−12.71

CALENDAR DATE	ROTATION NUMBER	HELIOGRAPHIC			DIAMETER	RA	DEC
		Lo	Bo	Po	(arcmin)	(HH:MM)	(deg)
2/16/2009	2080	284.72	–6.88	–17.77	32.377	21:59.0	–12.36
2/17/2009	2080	271.54	–6.91	–18.09	32.370	22: 2.9	–12.02
2/18/2009	2080	258.37	–6.95	–18.42	32.363	22: 6.7	–11.66
2/19/2009	2080	245.20	–6.99	–18.73	32.357	22:10.6	–11.31
2/20/2009	2080	232.03	–7.02	–19.04	32.350	22:14.4	–10.95
2/21/2009	2080	218.86	–7.05	–19.34	32.342	22:18.3	–10.59
2/22/2009	2080	205.68	–7.08	–19.64	32.335	22:22.1	–10.23
2/23/2009	2080	192.51	–7.11	–19.93	32.328	22:25.9	–9.86
2/24/2009	2080	179.34	–7.13	–20.22	32.320	22:29.7	–9.50
2/25/2009	2080	166.16	–7.15	–20.50	32.313	22:33.5	–9.13
2/26/2009	2080	152.99	–7.17	–20.77	32.305	22:37.3	–8.75
2/27/2009	2080	139.81	–7.19	–21.04	32.298	22:41.0	–8.38
2/28/2009	2080	126.70	–7.20	–21.30	32.290	22:44.8	–8.00
3/ 1/2009	2080	113.53	–7.22	–21.55	32.282	22:48.6	–7.62
3/ 2/2009	2080	100.35	–7.23	–21.80	32.274	22:52.3	–7.24
3/ 3/2009	2080	87.17	–7.24	–22.04	32.266	22:56.0	–6.86
3/ 4/2009	2080	74.00	–7.24	–22.28	32.258	22:59.8	–6.47
3/ 5/2009	2080	60.82	–7.25	–22.51	32.250	23: 3.5	–6.09
3/ 6/2009	2080	47.64	–7.25	–22.73	32.241	23: 7.2	–5.70
3/ 7/2009	2080	34.46	–7.25	–22.95	32.233	23:10.9	–5.31
3/ 8/2009	2080	21.28	–7.25	–23.16	32.225	23:14.6	–4.92
3/ 9/2009	2080	8.11	–7.24	–23.36	32.216	23:18.3	–4.53
3/10/2009	2081	354.93	–7.24	–23.56	32.208	23:22.0	–4.14
3/11/2009	2081	341.75	–7.23	–23.75	32.199	23:25.7	–3.75
3/12/2009	2081	328.56	–7.21	–23.93	32.190	23:29.4	–3.36
3/13/2009	2081	315.38	–7.20	–24.11	32.182	23:33.0	–2.96
3/14/2009	2081	302.20	–7.19	–24.28	32.173	23:36.7	–2.57
3/15/2009	2081	289.02	–7.17	–24.44	32.164	23:40.4	–2.17
3/16/2009	2081	275.84	–7.15	–24.60	32.155	23:44.0	–1.78
3/17/2009	2081	262.65	–7.12	–24.75	32.146	23:47.7	–1.38
3/18/2009	2081	249.47	–7.10	–24.89	32.137	23:51.4	–0.99
3/19/2009	2081	236.28	–7.07	–25.03	32.128	23:55.0	–0.59
3/20/2009	2081	223.10	–7.05	–25.16	32.119	23:58.7	–0.19
3/21/2009	2081	209.97	–7.01	–25.28	32.110	00: 2.3	0.20
3/22/2009	2081	196.78	–6.98	–25.39	32.101	00: 5.9	0.60
3/23/2009	2081	183.60	–6.95	–25.50	32.092	00: 9.6	0.99
3/24/2009	2081	170.41	–6.91	–25.60	32.083	00:13.2	1.39
3/25/2009	2081	157.22	–6.87	–25.70	32.074	00:16.9	1.78
3/26/2009	2081	144.03	–6.83	–25.79	32.065	00:20.5	2.17
3/27/2009	2081	130.84	–6.79	–25.87	32.055	00:24.2	2.56
3/28/2009	2081	117.65	–6.74	–25.94	32.046	00:27.8	2.96
3/29/2009	2081	104.45	–6.69	–26.01	32.037	00:31.4	3.35
3/30/2009	2081	91.26	–6.64	–26.07	32.028	00:35.1	3.74
3/31/2009	2081	78.07	–6.59	–26.12	32.019	00:38.7	4.12

(continued)

CALENDAR DATE	ROTATION NUMBER	HELIOGRAPHIC			DIAMETER	RA	DEC
		Lo	Bo	Po	(arcmin)	(HH:MM)	(deg)
4/ 1/2009	2081	64.87	−6.54	−26.16	32.009	00:42.4	4.51
4/ 2/2009	2081	51.68	−6.49	−26.20	32.000	00:46.0	4.90
4/ 3/2009	2081	38.48	−6.43	−26.23	31.991	00:49.7	5.28
4/ 4/2009	2081	25.28	−6.37	−26.26	31.982	00:53.3	5.66
4/ 5/2009	2081	12.09	−6.31	−26.27	31.973	00:57.0	6.04
4/ 6/2009	2082	358.89	−6.25	−26.28	31.963	01:0.6	6.42
4/ 7/2009	2082	345.69	−6.18	−26.28	31.954	01: 4.3	6.80
4/ 8/2009	2082	332.49	−6.12	−26.28	31.945	01: 8.0	7.18
4/ 9/2009	2082	319.29	−6.05	−26.26	31.936	01:11.6	7.55
4/10/2009	2082	306.15	−5.98	−26.24	31.927	01:15.3	7.92
4/11/2009	2082	292.94	−5.91	−26.21	31.918	01:19.0	8.29
4/12/2009	2082	279.74	−5.83	−26.18	31.909	01:22.7	8.66
4/13/2009	2082	266.54	−5.76	−26.14	31.900	01:26.4	9.02
4/14/2009	2082	253.33	−5.68	−26.09	31.891	01:30.0	9.38
4/15/2009	2082	240.13	−5.61	−26.03	31.882	01:33.7	9.74
4/16/2009	2082	226.92	−5.53	−25.97	31.873	01:37.5	10.10
4/17/2009	2082	213.71	−5.45	−25.89	31.864	01:41.2	10.45
4/18/2009	2082	200.50	−5.36	−25.82	31.855	01:44.9	10.80
4/19/2009	2082	187.29	−5.28	−25.73	31.846	01:48.6	11.15
4/20/2009	2082	174.08	−5.19	−25.64	31.838	01:52.3	11.50
4/21/2009	2082	160.87	−5.11	−25.53	31.829	01:56.1	11.84
4/22/2009	2082	147.66	−5.02	−25.43	31.820	01:59.8	12.18
4/23/2009	2082	134.45	−4.93	−25.31	31.812	02: 3.6	12.51
4/24/2009	2082	121.24	−4.84	−25.19	31.803	02: 7.3	12.84
4/25/2009	2082	108.02	−4.74	−25.06	31.795	02:11.1	13.17
4/26/2009	2082	94.81	−4.65	−24.92	31.786	02:14.9	13.49
4/27/2009	2082	81.59	−4.56	−24.77	31.778	02:18.6	13.82
4/28/2009	2082	68.37	−4.46	−24.62	31.770	02:22.4	14.13
4/29/2009	2082	55.16	−4.36	−24.46	31.762	02:26.2	14.45
4/30/2009	2082	42.00	−4.26	−24.29	31.754	02:30.0	14.75
5/ 1/2009	2082	28.78	−4.16	−24.12	31.746	02:33.9	15.06
5/ 2/2009	2082	15.56	−4.06	−23.94	31.738	02:37.7	15.36
5/ 3/2009	2082	2.34	−3.96	−23.75	31.730	02:41.5	15.66
5/ 4/2009	2083	349.12	−3.86	−23.55	31.722	02:45.4	15.95
5/ 5/2009	2083	335.90	−3.75	−23.35	31.714	02:49.2	16.24
5/ 6/2009	2083	322.68	−3.65	−23.14	31.706	02:53.1	16.52
5/ 7/2009	2083	309.45	−3.54	−22.92	31.699	02:57.0	16.80
5/ 8/2009	2083	296.23	−3.43	−22.70	31.691	03:0.8	17.07
5/ 9/2009	2083	283.01	−3.33	−22.47	31.684	03: 4.7	17.34
5/10/2009	2083	269.78	−3.22	−22.23	31.677	03: 8.6	17.61
5/11/2009	2083	256.56	−3.11	−21.98	31.670	03:12.5	17.87
5/12/2009	2083	243.33	−3.00	−21.73	31.663	03:16.5	18.12
5/13/2009	2083	230.10	−2.88	−21.47	31.656	03:20.4	18.37
5/14/2009	2083	216.87	−2.77	−21.21	31.649	03:24.3	18.61
5/15/2009	2083	203.65	−2.66	−20.94	31.642	03:28.3	18.85

CALENDAR DATE	ROTATION NUMBER	HELIOGRAPHIC			DIAMETER	RA	DEC
		Lo	Bo	Po	(arcmin)	(HH:MM)	(deg)
5/16/2009	2083	190.42	−2.55	−20.66	31.635	03:32.2	19.09
5/17/2009	2083	177.19	−2.43	−20.38	31.629	03:36.2	19.32
5/18/2009	2083	163.96	−2.32	−20.09	31.622	03:40.2	19.54
5/19/2009	2083	150.73	−2.20	−19.79	31.616	03:44.2	19.76
5/20/2009	2083	137.56	−2.09	−19.48	31.610	03:48.1	19.97
5/21/2009	2083	124.32	−1.97	−19.17	31.603	03:52.2	20.17
5/22/2009	2083	111.09	−1.85	−18.86	31.597	03:56.2	20.37
5/23/2009	2083	97.86	−1.73	−18.54	31.591	03:60.2	20.57
5/24/2009	2083	84.63	−1.62	−18.21	31.586	04: 4.2	20.76
5/25/2009	2083	71.39	−1.50	−17.88	31.580	04: 8.2	20.94
5/26/2009	2083	58.16	−1.38	−17.54	31.574	04:12.3	21.12
5/27/2009	2083	44.92	−1.26	−17.19	31.569	04:16.3	21.29
5/28/2009	2083	31.69	−1.14	−16.84	31.564	04:20.4	21.45
5/29/2009	2083	18.45	−1.02	−16.49	31.558	04:24.5	21.61
5/30/2009	2083	5.22	−0.90	−16.13	31.553	04:28.5	21.76
5/31/2009	2084	351.98	−0.78	−15.76	31.548	04:32.6	21.91
6/ 1/2009	2084	338.74	−0.66	−15.39	31.544	04:36.7	22.05
6/ 2/2009	2084	325.51	−0.54	−15.02	31.539	04:40.8	22.18
6/ 3/2009	2084	312.27	−0.42	−14.64	31.534	04:44.9	22.31
6/ 4/2009	2084	299.03	−0.30	−14.25	31.530	04:49.0	22.43
6/ 5/2009	2084	285.79	−0.18	−13.86	31.526	04:53.1	22.54
6/ 6/2009	2084	272.55	−0.06	13.47	31.522	04:57.3	22.65
6/ 7/2009	2084	259.31	0.06	−13.07	31.517	05: 1.4	22.75
6/ 8/2009	2084	246.08	0.18	−12.67	31.514	05: 5.5	22.84
6/ 9/2009	2084	232.90	0.31	−12.26	31.510	05: 9.6	22.93
6/10/2009	2084	219.66	0.43	−11.85	31.506	05:13.8	23.01
6/11/2009	2084	206.42	0.55	−11.44	31.503	05:17.9	23.08
6/12/2009	2084	193.18	0.67	−11.02	31.499	05:22.1	23.15
6/13/2009	2084	179.94	0.79	−10.60	31.496	05:26.2	23.21
6/14/2009	2084	166.70	0.91	−10.18	31.493	05:30.4	23.26
6/15/2009	2084	153.46	1.03	−9.75	31.490	05:34.5	23.31
6/16/2009	2084	140.22	1.14	−9.32	31.487	05:38.7	23.35
6/17/2009	2084	126.98	1.26	−8.89	31.485	05:42.8	23.38
6/18/2009	2084	113.74	1.38	−8.46	31.482	05:47.0	23.41
6/19/2009	2084	100.50	1.50	−8.02	31.480	05:51.1	23.42
6/20/2009	2084	87.26	1.62	−7.58	31.478	05:55.3	23.44
6/21/2009	2084	74.02	1.73	−7.14	31.476	05:59.5	23.44
6/22/2009	2084	60.78	1.85	−6.70	31.474	06: 3.6	23.44
6/23/2009	2084	47.54	1.97	−6.25	31.472	06: 7.8	23.43
6/24/2009	2084	34.29	2.08	−5.81	31.470	06:11.9	23.42
6/25/2009	2084	21.05	2.20	−5.36	31.469	06:16.1	23.39
6/26/2009	2084	7.81	2.31	−4.91	31.467	06:20.2	23.36
6/27/2009	2085	354.57	2.43	−4.46	31.466	06:24.4	23.33
6/28/2009	2085	341.33	2.54	−4.01	31.465	06:28.5	23.28
6/29/2009	2085	328.15	2.65	−3.55	31.464	06:32.7	23.24
6/30/2009	2085	314.91	2.76	−3.10	31.463	06:36.8	23.18

Appendix C

CALENDAR DATE	ROTATION NUMBER	HELIOGRAPHIC			DIAMETER	RA	DEC
		Lo	Bo	Po	(arcmin)	(HH:MM)	(deg)
7/ 1/2009	2085	301.67	2.87	−2.65	31.463	06:40.9	23.12
7/ 2/2009	2085	288.43	2.98	−2.20	31.462	06:45.1	23.05
7/ 3/2009	2085	275.19	3.09	−1.74	31.462	06:49.2	22.97
7/ 4/2009	2085	261.95	3.20	−1.29	31.462	06:53.3	22.89
7/ 5/2009	2085	248.72	3.31	−0.84	31.462	06:57.4	22.80
7/ 6/2009	2085	235.48	3.42	−0.38	31.462	07: 1.6	22.70
7/ 7/2009	2085	222.24	3.52	0.07	31.462	07: 5.7	22.60
7/ 8/2009	2085	209.00	3.63	0.52	31.463	07: 9.8	22.49
7/ 9/2009	2085	195.76	3.73	0.97	31.463	07:13.8	22.37
7/10/2009	2085	182.52	3.83	1.42	31.464	07:17.9	22.25
7/11/2009	2085	169.28	3.93	1.87	31.465	07:22.0	22.12
7/12/2009	2085	156.05	4.03	2.32	31.466	07:26.1	21.98
7/13/2009	2085	142.81	4.13	2.77	31.467	07:30.1	21.84
7/14/2009	2085	129.57	4.23	3.21	31.468	07:34.2	21.69
7/15/2009	2085	116.34	4.33	3.66	31.470	07:38.3	21.54
7/16/2009	2085	103.10	4.43	4.10	31.471	07:42.3	21.38
7/17/2009	2085	89.86	4.52	4.54	31.473	07:46.3	21.21
7/18/2009	2085	76.63	4.61	4.98	31.475	07:50.3	21.04
7/19/2009	2085	63.45	4.71	5.41	31.477	07:54.4	20.86
7/20/2009	2085	50.22	4.80	5.85	31.479	07:58.4	20.68
7/21/2009	2085	36.99	4.89	6.28	31.482	08: 2.4	20.49
7/22/2009	2085	23.75	4.98	6.71	31.484	08: 6.3	20.30
7/23/2009	2085	10.52	5.06	7.13	31.487	08:10.3	20.09
7/24/2009	2086	357.28	5.15	7.56	31.489	08:14.3	19.89
7/25/2009	2086	344.05	5.23	7.98	31.492	08:18.2	19.68
7/26/2009	2086	330.82	5.32	8.40	31.495	08:22.2	19.46
7/27/2009	2086	317.59	5.40	8.81	31.498	08:26.1	19.24
7/28/2009	2086	304.36	5.48	9.23	31.502	08:30.0	19.01
7/29/2009	2086	291.13	5.56	9.63	31.505	08:34.0	18.77
7/30/2009	2086	277.89	5.63	10.04	31.509	08:37.9	18.54
7/31/2009	2086	264.67	5.71	10.44	31.513	08:41.8	18.29
8/ 1/2009	2086	251.44	5.78	10.84	31.516	08:45.7	18.04
8/ 2/2009	2086	238.21	5.85	11.24	31.520	08:49.5	17.79
8/ 3/2009	2086	224.98	5.93	11.63	31.525	08:53.4	17.53
8/ 4/2009	2086	211.75	5.99	12.02	31.529	08:57.3	17.27
8/ 5/2009	2086	198.52	6.06	12.40	31.533	09: 1.1	17.00
8/ 6/2009	2086	185.30	6.13	12.78	31.538	09: 4.9	16.73
8/ 7/2009	2086	172.07	6.19	13.16	31.542	09: 8.8	16.45
8/ 8/2009	2086	158.84	6.25	13.53	31.547	09:12.6	16.17
8/ 9/2009	2086	145.68	6.31	13.90	31.552	09:16.4	15.88
8/10/2009	2086	132.46	6.37	14.26	31.557	09:20.2	15.59
8/11/2009	2086	119.23	6.43	14.62	31.562	09:24.0	15.30
8/12/2009	2086	106.01	6.49	14.98	31.568	09:27.8	15.00
8/13/2009	2086	92.79	6.54	15.33	31.573	09:31.5	14.70
8/14/2009	2086	79.56	6.59	15.67	31.578	09:35.3	14.39
8/15/2009	2086	66.34	6.64	16.02	31.584	09:39.0	14.08

CALENDAR DATE	ROTATION NUMBER	HELIOGRAPHIC			DIAMETER	RA	DEC
		Lo	Bo	Po	(arcmin)	(HH:MM)	(deg)
8/16/2009	2086	53.12	6.69	16.35	31.590	09:42.8	13.77
8/17/2009	2086	39.90	6.73	16.69	31.596	09:46.5	13.45
8/18/2009	2086	26.68	6.78	17.01	31.602	09:50.2	13.13
8/19/2009	2086	13.46	6.82	17.34	31.608	09:54.0	12.81
8/20/2009	2086	0.24	6.86	17.66	31.614	09:57.7	12.48
8/21/2009	2087	347.02	6.90	17.97	31.620	10: 1.4	12.15
8/22/2009	2087	333.80	6.94	18.28	31.627	10: 5.1	11.82
8/23/2009	2087	320.59	6.97	18.58	31.633	10: 8.8	11.48
8/24/2009	2087	307.37	7.00	18.88	31.640	10:12.4	11.14
8/25/2009	2087	294.15	7.03	19.17	31.647	10:16.1	10.79
8/26/2009	2087	280.94	7.06	19.46	31.654	10:19.8	10.45
8/27/2009	2087	267.72	7.09	19.74	31.661	10:23.4	10.10
8/28/2009	2087	254.51	7.11	20.02	31.668	10:27.1	9.75
8/29/2009	2087	241.36	7.14	20.29	31.675	10:30.7	9.39
8/30/2009	2087	228.14	7.16	20.56	31.682	10:34.4	9.04
8/31/2009	2087	214.93	7.18	20.82	31.689	10:38.0	8.68
9/ 1/2009	2087	201.72	7.19	21.08	31.697	10:41.6	8.32
9/ 2/2009	2087	188.51	7.21	21.33	31.704	10:45.3	7.95
9/ 3/2009	2087	175.30	7.22	21.57	31.712	10:48.9	7.59
9/ 4/2009	2087	162.09	7.23	21.81	31.720	10:52.5	7.22
9/ 5/2009	2087	148.87	7.24	22.05	31.728	10:56.1	6.85
9/ 6/2009	2087	135.67	7.24	22.28	31.735	10:59.7	6.48
9/ 7/2009	2087	122.46	7.25	22.50	31.743	11: 3.3	6.10
9/ 8/2009	2087	109.25	7.25	22.71	31.751	11: 6.9	5.73
9/ 9/2009	2087	96.04	7.25	22.92	31.759	11:10.5	5.35
9/10/2009	2087	82.83	7.25	23.13	31.768	11:14.1	4.97
9/11/2009	2087	69.63	7.24	23.33	31.776	11:17.7	4.59
9/12/2009	2087	56.42	7.24	23.52	31.784	11:21.3	4.21
9/13/2009	2087	43.21	7.23	23.71	31.793	11:24.9	3.83
9/14/2009	2087	30.01	7.22	23.89	31.801	11:28.5	3.45
9/15/2009	2087	16.80	7.20	24.06	31.809	11:32.1	3.06
9/16/2009	2087	3.60	7.19	24.23	31.818	11:35.7	2.68
9/17/2009	2088	350.39	7.17	24.39	31.827	11:39.3	2.29
9/18/2009	2088	337.25	7.15	24.55	31.835	11:42.9	1.91
9/19/2009	2088	324.05	7.13	24.70	31.844	11:46.4	1.52
9/20/2009	2088	310.85	7.11	24.84	31.853	11:50.0	1.13
9/21/2009	2088	297.64	7.08	24.98	31.862	11:53.6	0.74
9/22/2009	2088	284.44	7.06	25.11	31.870	11:57.2	0.35
9/23/2009	2088	271.24	7.03	25.23	31.879	12:0.8	−0.04
9/24/2009	2088	258.04	7.00	25.35	31.888	12: 4.4	−0.43
9/25/2009	2088	244.84	6.96	25.46	31.897	12: 8.0	−0.82
9/26/2009	2088	231.64	6.93	25.56	31.906	12:11.6	−1.21
9/27/2009	2088	218.44	6.89	25.66	31.915	12:15.2	−1.60
9/28/2009	2088	205.24	6.85	25.75	31.924	12:18.8	−1.99
9/29/2009	2088	192.04	6.81	25.83	31.933	12:22.4	−2.38
9/30/2009	2088	178.84	6.76	25.91	31.943	12:26.0	−2.77

CALENDAR DATE	ROTATION NUMBER	HELIOGRAPHIC			DIAMETER	RA	DEC
		Lo	Bo	Po	(arcmin)	(HH:MM)	(deg)
10/ 1/2009	2088	165.65	6.72	25.97	31.952	12:29.7	–3.15
10/ 2/2009	2088	152.45	6.67	26.04	31.961	12:33.3	–3.54
10/ 3/2009	2088	139.25	6.62	26.09	31.970	12:36.9	–3.93
10/ 4/2009	2088	126.05	6.57	26.14	31.979	12:40.5	–4.32
10/ 5/2009	2088	112.86	6.51	26.18	31.988	12:44.2	–4.70
10/ 6/2009	2088	99.66	6.46	26.22	31.998	12:47.8	–5.09
10/ 7/2009	2088	86.46	6.40	26.24	32.007	12:51.5	–5.47
10/ 8/2009	2088	73.33	6.34	26.26	32.016	12:55.1	–5.85
10/ 9/2009	2088	60.14	6.28	26.28	32.025	12:58.8	–6.23
10/10/2009	2088	46.94	6.21	26.28	32.034	13: 2.5	–6.61
10/11/2009	2088	33.75	6.15	26.28	32.044	13: 6.2	–6.99
10/12/2009	2088	20.55	6.08	26.27	32.053	13: 9.9	–7.37
10/13/2009	2088	7.36	6.01	26.25	32.062	13:13.6	–7.74
10/14/2009	2089	354.17	5.94	26.23	32.071	13:17.3	–8.12
10/15/2009	2089	340.97	5.87	26.20	32.080	13:21.0	–8.49
10/16/2009	2089	327.78	5.79	26.16	32.090	13:24.7	–8.86
10/17/2009	2089	314.59	5.72	26.11	32.099	13:28.4	–9.23
10/18/2009	2089	301.39	5.64	26.06	32.108	13:32.2	–9.59
10/19/2009	2089	288.20	5.56	25.99	32.117	13:35.9	–9.95
10/20/2009	2089	275.01	5.48	25.92	32.126	13:39.7	–10.31
10/21/2009	2089	261.82	5.39	25.85	32.135	13:43.5	–10.67
10/22/2009	2089	248.63	5.31	25.76	32.144	13:47.3	–11.03
10/23/2009	2089	235.43	5.22	25.67	32.153	13:51.1	–11.38
10/24/2009	2089	222.24	5.14	25.57	32.162	13:54.9	–11.73
10/25/2009	2089	209.05	5.05	25.46	32.170	13:58.7	–12.08
10/26/2009	2089	195.86	4.95	25.34	32.179	14: 2.5	–12.42
10/27/2009	2089	182.67	4.86	25.22	32.188	14: 6.4	–12.76
10/28/2009	2089	169.55	4.77	25.08	32.197	14:10.2	–13.10
10/29/2009	2089	156.36	4.67	24.94	32.205	14:14.1	–13.43
10/30/2009	2089	143.17	4.57	24.80	32.214	14:18.0	–13.76
10/31/2009	2089	129.98	4.47	24.64	32.222	14:21.9	–14.09
11/ 1/2009	2089	116.79	4.37	24.48	32.231	14:25.8	–14.41
11/ 2/2009	2089	103.60	4.27	24.31	32.239	14:29.7	–14.73
11/ 3/2009	2089	90.41	4.17	24.13	32.247	14:33.7	–15.04
11/ 4/2009	2089	77.22	4.07	23.94	32.256	14:37.6	–15.35
11/ 5/2009	2089	64.04	3.96	23.75	32.264	14:41.6	–15.66
11/ 6/2009	2089	50.85	3.85	23.54	32.272	14:45.6	–15.96
11/ 7/2009	2089	37.66	3.74	23.33	32.280	14:49.5	–16.26
11/ 8/2009	2089	24.47	3.64	23.11	32.288	14:53.5	–16.55
11/ 9/2009	2089	11.29	3.53	22.89	32.295	14:57.6	–16.84
11/10/2009	2090	358.10	3.41	22.65	32.303	15: 1.6	–17.13
11/11/2009	2090	344.91	3.30	22.41	32.311	15: 5.6	–17.40
11/12/2009	2090	331.73	3.19	22.16	32.318	15: 9.7	–17.68
11/13/2009	2090	318.54	3.07	21.91	32.326	15:13.8	–17.95
11/14/2009	2090	305.35	2.96	21.64	32.333	15:17.9	–18.21
11/15/2009	2090	292.17	2.84	21.37	32.340	15:22.0	–18.47

CALENDAR DATE	ROTATION NUMBER	HELIOGRAPHIC			DIAMETER	RA	DEC
		Lo	Bo	Po	(arcmin)	(HH:MM)	(deg)
11/16/2009	2090	278.98	2.72	21.09	32.348	15:26.1	−18.72
11/17/2009	2090	265.86	2.60	20.80	32.355	15:30.2	−18.97
11/18/2009	2090	252.67	2.49	20.51	32.362	15:34.4	−19.21
11/19/2009	2090	239.49	2.37	20.21	32.368	15:38.5	−19.45
11/20/2009	2090	226.30	2.24	19.90	32.375	15:42.7	−19.68
11/21/2009	2090	213.12	2.12	19.58	32.382	15:46.9	−19.90
11/22/2009	2090	199.94	2.00	19.26	32.388	15:51.1	−20.12
11/23/2009	2090	186.75	1.88	18.93	32.395	15:55.3	−20.33
11/24/2009	2090	173.57	1.75	18.59	32.401	15:59.5	−20.54
11/25/2009	2090	160.38	1.63	18.25	32.407	16: 3.7	−20.74
11/26/2009	2090	147.20	1.51	17.90	32.413	16: 8.0	−20.93
11/27/2009	2090	134.02	1.38	17.54	32.419	16:12.3	−21.12
11/28/2009	2090	120.84	1.25	17.18	32.425	16:16.5	−21.30
11/29/2009	2090	107.65	1.13	16.81	32.430	16:20.8	−21.47
11/30/2009	2090	94.47	1.00	16.43	32.436	16:25.1	−21.63
12/ 1/2009	2090	81.29	0.88	16.05	32.441	16:29.4	−21.79
12/ 2/2009	2090	68.11	0.75	15.66	32.446	16:33.7	−21.95
12/ 3/2009	2090	54.92	0.62	15.27	32.451	16:38.1	−22.09
12/ 4/2009	2090	41.74	0.49	14.87	32.456	16:42.4	−22.23
12/ 5/2009	2090	28.56	0.37	14.46	32.461	16:46.8	−22.36
12/ 6/2009	2090	15.38	0.24	14.05	32.465	16:51.1	−22.49
12/ 7/2009	2090	2.26	0.11	13.64	32.470	16:55.5	−22.60
12/ 8/2009	2091	349.08	−0.02	13.22	32.474	16:59.9	−22.71
12/ 9/2009	2091	335.90	−0.15	12.79	32.478	17: 4.2	−22.81
12/10/2009	2091	322.72	−0.27	12.36	32.483	17: 8.6	−22.91
12/11/2009	2091	309.54	−0.40	11.93	32.486	17:13.0	−22.99
12/12/2009	2091	296.36	−0.53	11.49	32.490	17:17.4	−23.07
12/13/2009	2091	283.18	−0.66	11.05	32.494	17:21.8	−23.15
12/14/2009	2091	270.00	−0.79	10.60	32.497	17:26.3	−23.21
12/15/2009	2091	256.83	−0.91	10.15	32.500	17:30.7	−23.27
12/16/2009	2091	243.65	−1.04	9.69	32.503	17:35.1	−23.31
12/17/2009	2091	230.47	−1.17	9.23	32.506	17:39.5	−23.35
12/18/2009	2091	217.29	−1.29	8.77	32.509	17:44.0	−23.39
12/19/2009	2091	204.11	−1.42	8.31	32.512	17:48.4	−23.41
12/20/2009	2091	190.94	−1.55	7.84	32.514	17:52.8	−23.43
12/21/2009	2091	177.76	−1.67	7.37	32.517	17:57.3	−23.44
12/22/2009	2091	164.58	−1.80	6.90	32.519	18: 1.7	−23.44
12/23/2009	2091	151.40	−1.92	6.43	32.521	18: 6.1	−23.44
12/24/2009	2091	138.23	−2.04	5.95	32.523	18:10.6	−23.42
12/25/2009	2091	125.05	−2.17	5.47	32.524	18:15.0	−23.40
12/26/2009	2091	111.88	−2.29	4.99	32.526	18:19.4	−23.37
12/27/2009	2091	98.70	−2.41	4.51	32.527	18:23.9	−23.33
12/28/2009	2091	85.59	−2.53	4.03	32.528	18:28.3	−23.29
12/29/2009	2091	72.41	−2.65	3.55	32.529	18:32.7	−23.23
12/30/2009	2091	59.24	−2.77	3.06	32.530	18:37.2	−23.17
12/31/2009	2091	46.06	−2.89	2.58	32.530	18:41.6	−23.11

CALENDAR DATE	ROTATION NUMBER	HELIOGRAPHIC			DIAMETER	RA	DEC
		Lo	Bo	Po	(arcmin)	(HH:MM)	(deg)
1/ 1/2010	2091	32.89	−3.01	2.10	32.531	18:46.0	−23.03
1/ 2/2010	2091	19.71	−3.12	1.61	32.531	18:50.4	−22.95
1/ 3/2010	2091	6.54	−3.24	1.13	32.531	18:54.8	−22.85
1/ 4/2010	2092	353.37	−3.35	0.64	32.531	18:59.2	−22.75
1/ 5/2010	2092	340.19	−3.47	0.16	32.531	19: 3.6	−22.65
1/ 6/2010	2092	327.02	−3.58	−0.33	32.531	19: 8.0	−22.53
1/ 7/2010	2092	313.85	−3.69	−0.81	32.530	19:12.4	−22.41
1/ 8/2010	2092	300.67	−3.80	−1.29	32.529	19:16.7	−22.28
1/ 9/2010	2092	287.50	−3.91	−1.77	32.529	19:21.1	−22.15
1/10/2010	2092	274.33	−4.02	−2.25	32.528	19:25.4	−22.00
1/11/2010	2092	261.16	−4.12	−2.73	32.526	19:29.8	−21.85
1/12/2010	2092	247.98	−4.23	−3.20	32.525	19:34.1	−21.70
1/13/2010	2092	234.81	−4.33	−3.68	32.523	19:38.4	−21.53
1/14/2010	2092	221.64	−4.44	−4.15	32.522	19:42.8	−21.36
1/15/2010	2092	208.47	−4.54	−4.62	32.520	19:47.1	−21.18
1/16/2010	2092	195.30	−4.64	−5.09	32.518	19:51.4	−21.00
1/17/2010	2092	182.19	−4.74	−5.55	32.516	19:55.6	−20.81
1/18/2010	2092	169.02	−4.83	−6.01	32.513	19:59.9	−20.61
1/19/2010	2092	155.85	−4.93	−6.47	32.511	20: 4.2	−20.40
1/20/2010	2092	142.68	−5.02	−6.93	32.508	20: 8.4	−20.19
1/21/2010	2092	129.51	−5.11	−7.38	32.505	20:12.6	−19.97
1/22/2010	2092	116.34	−5.20	−7.83	32.502	20:16.9	−19.75
1/23/2010	2092	103.17	−5.29	−8.28	32.499	20:21.1	−19.52
1/24/2010	2092	90.00	−5.38	−8.72	32.495	20:25.3	−19.28
1/25/2010	2092	76.82	−5.46	−9.16	32.492	20:29.4	−19.04
1/26/2010	2092	63.66	−5.55	−9.60	32.488	20:33.6	−18.80
1/27/2010	2092	50.49	−5.63	−10.03	32.485	20:37.8	−18.54
1/28/2010	2092	37.32	−5.71	−10.46	32.481	20:41.9	−18.28
1/29/2010	2092	24.15	−5.79	−10.88	32.476	20:46.0	−18.02
1/30/2010	2092	10.98	−5.87	−11.30	32.472	20:50.1	−17.75
1/31/2010	2093	357.81	−5.94	−11.71	32.468	20:54.2	−17.47
2/ 1/2010	2093	344.64	−6.01	−12.12	32.463	20:58.3	−17.19
2/ 2/2010	2093	331.47	−6.08	−12.53	32.459	21: 2.4	−16.91
2/ 3/2010	2093	318.30	−6.15	−12.93	32.454	21: 6.5	−16.62
2/ 4/2010	2093	305.13	−6.22	−13.33	32.449	21:10.5	−16.32
2/ 5/2010	2093	291.96	−6.28	−13.72	32.444	21:14.5	−16.02
2/ 6/2010	2093	278.85	−6.35	−14.10	32.438	21:18.6	−15.72
2/ 7/2010	2093	265.68	−6.41	−14.49	32.433	21:22.6	−15.41
2/ 8/2010	2093	252.51	−6.47	−14.86	32.427	21:26.6	−15.10
2/ 9/2010	2093	239.34	−6.52	−15.23	32.422	21:30.5	−14.78
2/10/2010	2093	226.17	−6.58	−15.60	32.416	21:34.5	−14.46
2/11/2010	2093	213.00	−6.63	−15.96	32.410	21:38.5	−14.13
2/12/2010	2093	199.83	−6.68	−16.32	32.404	21:42.4	−13.80
2/13/2010	2093	186.66	−6.73	−16.67	32.398	21:46.3	−13.47
2/14/2010	2093	173.49	−6.78	−17.01	32.391	21:50.2	−13.13
2/15/2010	2093	160.31	−6.82	−17.35	32.385	21:54.1	−12.79

CALENDAR DATE	ROTATION NUMBER	HELIOGRAPHIC			DIAMETER	RA	DEC
		Lo	Bo	Po	(arcmin)	(HH:MM)	(deg)
2/16/2010	2093	147.14	–6.86	–17.69	32.379	21:58.0	–12.45
2/17/2010	2093	133.97	–6.90	–18.01	32.372	22: 1.9	–12.10
2/18/2010	2093	120.80	–6.94	–18.34	32.365	22: 5.8	–11.75
2/19/2010	2093	107.63	–6.98	–18.65	32.358	22: 9.6	–11.40
2/20/2010	2093	94.46	–7.01	–18.96	32.351	22:13.5	–11.04
2/21/2010	2093	81.28	–7.04	–19.27	32.344	22:17.3	–10.68
2/22/2010	2093	68.11	–7.07	–19.57	32.337	22:21.2	–10.32
2/23/2010	2093	54.94	–7.10	–19.86	32.330	22:25.0	–9.95
2/24/2010	2093	41.76	–7.12	–20.15	32.322	22:28.8	–9.58
2/25/2010	2093	28.59	–7.15	–20.43	32.315	22:32.6	–9.22
2/26/2010	2093	15.48	–7.17	–20.70	32.307	22:36.3	–8.84
2/27/2010	2093	2.30	–7.19	–20.97	32.299	22:40.1	–8.47
2/28/2010	2094	349.13	–7.20	–21.23	32.292	22:43.9	–8.09
3/ 1/2010	2094	335.95	–7.21	–21.49	32.284	22:47.6	–7.71
3/ 2/2010	2094	322.78	–7.23	–21.74	32.276	22:51.4	–7.33
3/ 3/2010	2094	309.60	–7.24	–21.98	32.268	22:55.1	–6.95
3/ 4/2010	2094	296.42	–7.24	–22.22	32.260	22:58.9	–6.57
3/ 5/2010	2094	283.25	–7.25	–22.45	32.252	23: 2.6	–6.18
3/ 6/2010	2094	270.07	–7.25	–22.68	32.243	23: 6.3	–5.79
3/ 7/2010	2094	256.89	–7.25	–22.89	32.235	23:10.0	–5.41
3/ 8/2010	2094	243.71	–7.25	–23.11	32.227	23:13.7	–5.02
3/ 9/2010	2094	230.53	–7.24	–23.31	32.218	23:17.4	–4.63
3/10/2010	2094	217.36	–7.24	–23.51	32.210	23:21.1	–4.24
3/11/2010	2094	204.18	–7.23	–23.70	32.201	23:24.8	–3.84
3/12/2010	2094	190.99	–7.22	–23.89	32.192	23:28.5	–3.45
3/13/2010	2094	177.81	–7.20	–24.06	32.184	23:32.2	–3.06
3/14/2010	2094	164.63	–7.19	–24.24	32.175	23:35.8	–2.66
3/15/2010	2094	151.45	–7.17	–24.40	32.166	23:39.5	–2.27
3/16/2010	2094	138.27	–7.15	–24.56	32.157	23:43.2	–1.87
3/17/2010	2094	125.08	–7.13	–24.71	32.148	23:46.8	–1.48
3/18/2010	2094	111.96	–7.11	–24.86	32.139	23:50.5	–1.08
3/19/2010	2094	98.78	–7.08	–24.99	32.130	23:54.1	–0.69
3/20/2010	2094	85.59	–7.05	–25.12	32.121	23:57.8	–0.29
3/21/2010	2094	72.40	–7.02	–25.25	32.112	00: 1.4	0.11
3/22/2010	2094	59.22	–6.99	–25.37	32.103	00: 5.1	0.50
3/23/2010	2094	46.03	–6.96	–25.48	32.094	00: 8.7	0.90
3/24/2010	2094	32.84	–6.92	–25.58	32.085	00:12.4	1.29
3/25/2010	2094	19.65	–6.88	–25.68	32.076	00:16.0	1.68
3/26/2010	2094	6.46	–6.84	–25.76	32.067	00:19.6	2.08
3/27/2010	2095	353.27	–6.80	–25.85	32.058	00:23.3	2.47
3/28/2010	2095	340.08	–6.75	–25.92	32.048	00:26.9	2.86
3/29/2010	2095	326.89	–6.71	–25.99	32.039	00:30.6	3.25
3/30/2010	2095	313.69	–6.66	–26.05	32.030	00:34.2	3.64
3/31/2010	2095	300.50	–6.61	–26.10	32.021	00:37.9	4.03

CALENDAR DATE	ROTATION NUMBER	HELIOGRAPHIC			DIAMETER	RA	DEC
		Lo	Bo	Po	(arcmin)	(HH:MM)	(deg)
4/ 1/2010	2095	287.31	−6.55	−26.15	32.012	00:41.5	4.42
4/ 2/2010	2095	274.11	−6.50	−26.19	32.002	00:45.1	4.80
4/ 3/2010	2095	260.91	−6.44	−26.22	31.993	00:48.8	5.19
4/ 4/2010	2095	247.72	−6.38	−26.25	31.984	00:52.4	5.57
4/ 5/2010	2095	234.52	−6.32	−26.27	31.975	00:56.1	5.95
4/ 6/2010	2095	221.32	−6.26	−26.28	31.966	00:59.8	6.33
4/ 7/2010	2095	208.18	−6.20	−26.28	31.956	01: 3.4	6.71
4/ 8/2010	2095	194.98	−6.13	−26.28	31.947	01: 7.1	7.09
4/ 9/2010	2095	181.78	−6.07	−26.27	31.938	01:10.7	7.46
4/10/2010	2095	168.58	−6.00	−26.25	31.929	01:14.4	7.83
4/11/2010	2095	155.38	−5.93	−26.22	31.920	01:18.1	8.20
4/12/2010	2095	142.18	−5.85	−26.19	31.911	01:21.8	8.57
4/13/2010	2095	128.97	−5.78	−26.15	31.902	01:25.5	8.93
4/14/2010	2095	115.77	−5.70	−26.10	31.893	01:29.2	9.30
4/15/2010	2095	102.56	−5.63	−26.04	31.884	01:32.9	9.65
4/16/2010	2095	89.35	−5.55	−25.98	31.875	01:36.6	10.01
4/17/2010	2095	76.15	−5.47	−25.91	31.866	01:40.3	10.37
4/18/2010	2095	62.94	−5.38	−25.83	31.857	01:44.0	10.72
4/19/2010	2095	49.73	−5.30	−25.75	31.848	01:47.7	11.07
4/20/2010	2095	36.52	−5.22	−25.66	31.840	01:51.4	11.41
4/21/2010	2095	23.31	−5.13	−25.56	31.831	01:55.2	11.75
4/22/2010	2095	10.10	−5.04	−25.45	31.822	01:58.9	12.09
4/23/2010	2096	356.89	−4.95	−25.34	31.814	02: 2.7	12.43
4/24/2010	2096	343.67	−4.86	−25.22	31.805	02: 6.4	12.76
4/25/2010	2096	330.46	−4.77	−25.09	31.797	02:10.2	13.09
4/26/2010	2096	317.24	−4.68	−24.95	31.788	02:14.0	13.42
4/27/2010	2096	304.09	−4.58	−24.81	31.780	02:17.7	13.74
4/28/2010	2096	290.88	−4.48	−24.66	31.772	02:21.5	14.06
4/29/2010	2096	277.66	−4.39	−24.50	31.764	02:25.3	14.37
4/30/2010	2096	264.44	−4.29	−24.33	31.755	02:29.1	14.68
5/ 1/2010	2096	251.22	−4.19	−24.16	31.747	02:32.9	14.99
5/ 2/2010	2096	238.00	−4.09	−23.98	31.739	02:36.8	15.29
5/ 3/2010	2096	224.78	−3.99	−23.79	31.732	02:40.6	15.59
5/ 4/2010	2096	211.56	−3.88	−23.60	31.724	02:44.4	15.88
5/ 5/2010	2096	198.34	−3.78	−23.40	31.716	02:48.3	16.17
5/ 6/2010	2096	185.12	−3.67	−23.19	31.708	02:52.2	16.45
5/ 7/2010	2096	171.89	−3.57	−22.97	31.701	02:56.0	16.73
5/ 8/2010	2096	158.67	−3.46	−22.75	31.693	02:59.9	17.01
5/ 9/2010	2096	145.45	−3.35	−22.52	31.686	03: 3.8	17.28
5/10/2010	2096	132.22	−3.24	−22.29	31.679	03: 7.7	17.54
5/11/2010	2096	119.00	−3.14	−22.04	31.671	03:11.6	17.80
5/12/2010	2096	105.77	−3.02	−21.79	31.664	03:15.5	18.06
5/13/2010	2096	92.54	−2.91	−21.54	31.657	03:19.4	18.31
5/14/2010	2096	79.31	−2.80	−21.27	31.650	03:23.4	18.56
5/15/2010	2096	66.09	−2.69	−21.00	31.644	03:27.3	18.80

CALENDAR DATE	ROTATION NUMBER	HELIOGRAPHIC			DIAMETER	RA	DEC
		Lo	Bo	Po	(arcmin)	(HH:MM)	(deg)
5/16/2010	2096	52.86	−2.58	−20.73	31.637	03:31.3	19.03
5/17/2010	2096	39.63	−2.46	−20.44	31.630	03:35.2	19.26
5/18/2010	2096	26.46	−2.35	−20.16	31.624	03:39.2	19.49
5/19/2010	2096	13.23	−2.23	−19.86	31.617	03:43.2	19.70
5/20/2010	2097	360.00	−2.12	−19.56	31.611	03:47.2	19.92
5/21/2010	2097	346.77	−2.00	−19.25	31.605	03:51.2	20.12
5/22/2010	2097	333.53	−1.88	−18.94	31.599	03:55.2	20.33
5/23/2010	2097	320.30	−1.76	−18.62	31.593	03:59.2	20.52
5/24/2010	2097	307.07	−1.65	−18.29	31.587	04: 3.2	20.71
5/25/2010	2097	293.83	−1.53	−17.96	31.581	04: 7.3	20.90
5/26/2010	2097	280.60	−1.41	−17.62	31.576	04:11.3	21.07
5/27/2010	2097	267.37	−1.29	−17.28	31.570	04:15.4	21.25
5/28/2010	2097	254.13	−1.17	−16.93	31.565	04:19.4	21.41
5/29/2010	2097	240.90	−1.05	−16.57	31.560	04:23.5	21.57
5/30/2010	2097	227.66	−0.93	−16.22	31.555	04:27.6	21.73
5/31/2010	2097	214.42	−0.81	−15.85	31.550	04:31.6	21.87
6/ 1/2010	2097	201.19	−0.69	−15.48	31.545	04:35.7	22.01
6/ 2/2010	2097	187.95	−0.57	−15.11	31.540	04:39.8	22.15
6/ 3/2010	2097	174.71	−0.45	−14.73	31.535	04:43.9	22.28
6/ 4/2010	2097	161.47	−0.33	−14.34	31.531	04:48.0	22.40
6/ 5/2010	2097	148.24	−0.21	−13.96	31.527	04:52.1	22.51
6/ 6/2010	2097	135.00	−0.09	−13.56	31.523	04:56.3	22.62
6/ 7/2010	2097	121.82	0.03	−13.17	31.518	04:60.4	22.72
6/ 8/2010	2097	108.58	0.15	−12.77	31.515	05: 4.5	22.82
6/ 9/2010	2097	95.34	0.27	−12.36	31.511	05: 8.6	22.91
6/10/2010	2097	82.10	0.39	−11.95	31.507	05:12.8	22.99
6/11/2010	2097	68.86	0.51	−11.54	31.504	05:16.9	23.06
6/12/2010	2097	55.62	0.64	−11.12	31.500	05:21.1	23.13
6/13/2010	2097	42.38	0.76	−10.70	31.497	05:25.2	23.19
6/14/2010	2097	29.14	0.87	−10.28	31.494	05:29.4	23.25
6/15/2010	2097	15.90	0.99	−9.86	31.491	05:33.5	23.30
6/16/2010	2097	2.66	1.11	−9.43	31.488	05:37.7	23.34
6/17/2010	2098	349.42	1.23	−9.00	31.485	05:41.8	23.37
6/18/2010	2098	336.18	1.35	−8.56	31.483	05:46.0	23.40
6/19/2010	2098	322.94	1.47	−8.13	31.480	05:50.1	23.42
6/20/2010	2098	309.70	1.59	−7.69	31.478	05:54.3	23.43
6/21/2010	2098	296.46	1.70	−7.25	31.476	05:58.5	23.44
6/22/2010	2098	283.22	1.82	−6.80	31.474	06: 2.6	23.44
6/23/2010	2098	269.98	1.94	−6.36	31.472	06: 6.8	23.43
6/24/2010	2098	256.74	2.05	−5.91	31.471	06:10.9	23.42
6/25/2010	2098	243.50	2.17	−5.47	31.469	06:15.1	23.40
6/26/2010	2098	230.26	2.28	−5.02	31.468	06:19.2	23.37
6/27/2010	2098	217.08	2.40	−4.57	31.466	06:23.4	23.34
6/28/2010	2098	203.84	2.51	−4.12	31.465	06:27.5	23.30
6/29/2010	2098	190.60	2.62	−3.67	31.464	06:31.7	23.25
6/30/2010	2098	177.36	2.73	−3.21	31.464	06:35.8	23.19

CALENDAR DATE	ROTATION NUMBER	HELIOGRAPHIC			DIAMETER	RA	DEC
		Lo	Bo	Po	(arcmin)	(HH:MM)	(deg)
7/ 1/2010	2098	164.12	2.84	−2.76	31.463	06:39.9	23.13
7/ 2/2010	2098	150.88	2.96	−2.31	31.462	06:44.1	23.06
7/ 3/2010	2098	137.64	3.06	−1.85	31.462	06:48.2	22.99
7/ 4/2010	2098	124.40	3.17	−1.40	31.462	06:52.3	22.91
7/ 5/2010	2098	111.16	3.28	−0.95	31.462	06:56.4	22.82
7/ 6/2010	2098	97.92	3.39	−0.49	31.462	07:0 .6	22.72
7/ 7/2010	2098	84.68	3.49	−0.04	31.462	07: 4.7	22.62
7/ 8/2010	2098	71.44	3.60	0.41	31.463	07: 8.8	22.51
7/ 9/2010	2098	58.20	3.70	0.86	31.463	07:12.9	22.40
7/10/2010	2098	44.97	3.81	1.31	31.464	07:16.9	22.28
7/11/2010	2098	31.73	3.91	1.76	31.465	07:21.0	22.15
7/12/2010	2098	18.49	4.01	2.21	31.466	07:25.1	22.02
7/13/2010	2098	5.25	4.11	2.66	31.467	07:29.2	21.88
7/14/2010	2099	352.02	4.21	3.10	31.468	07:33.2	21.73
7/15/2010	2099	338.78	4.30	3.55	31.469	07:37.3	21.58
7/16/2010	2099	325.54	4.40	3.99	31.471	07:41.3	21.42
7/17/2010	2099	312.37	4.50	4.43	31.473	07:45.3	21.25
7/18/2010	2099	299.13	4.59	4.87	31.475	07:49.4	21.08
7/19/2010	2099	285.90	4.68	5.31	31.477	07:53.4	20.91
7/20/2010	2099	272.66	4.77	5.74	31.479	07:57.4	20.73
7/21/2010	2099	259.43	4.86	6.17	31.481	08: 1.4	20.54
7/22/2010	2099	246.19	4.95	6.60	31.483	08: 5.4	20.34
7/23/2010	2099	232.96	5.04	7.03	31.486	08: 9.4	20.14
7/24/2010	2099	219.73	5.13	7.45	31.489	08:13.3	19.94
7/25/2010	2099	206.49	5.21	7.88	31.492	08:17.3	19.73
7/26/2010	2099	193.26	5.29	8.30	31.495	08:21.2	19.51
7/27/2010	2099	180.03	5.38	8.71	31.498	08:25.2	19.29
7/28/2010	2099	166.80	5.46	9.12	31.501	08:29.1	19.06
7/29/2010	2099	153.57	5.54	9.53	31.504	08:33.0	18.83
7/30/2010	2099	140.34	5.61	9.94	31.508	08:36.9	18.59
7/31/2010	2099	127.11	5.69	10.34	31.512	08:40.8	18.35
8/ 1/2010	2099	113.88	5.76	10.74	31.515	08:44.7	18.10
8/ 2/2010	2099	100.65	5.84	11.14	31.519	08:48.6	17.85
8/ 3/2010	2099	87.42	5.91	11.53	31.524	08:52.5	17.59
8/ 4/2010	2099	74.19	5.98	11.92	31.528	08:56.3	17.33
8/ 5/2010	2099	60.96	6.04	12.31	31.532	08:60.2	17.07
8/ 6/2010	2099	47.80	6.11	12.69	31.537	09: 4.0	16.79
8/ 7/2010	2099	34.57	6.17	13.06	31.541	09: 7.8	16.52
8/ 8/2010	2099	21.35	6.24	13.44	31.546	09:11.7	16.24
8/ 9/2010	2099	8.12	6.30	13.81	31.551	09:15.5	15.95
8/10/2010	2100	354.90	6.36	14.17	31.556	09:19.3	15.66
8/11/2010	2100	341.67	6.42	14.53	31.561	09:23.1	15.37
8/12/2010	2100	328.45	6.47	14.89	31.566	09:26.8	15.07
8/13/2010	2100	315.23	6.53	15.24	31.572	09:30.6	14.77
8/14/2010	2100	302.00	6.58	15.59	31.577	09:34.4	14.47
8/15/2010	2100	288.78	6.63	15.93	31.583	09:38.1	14.16

CALENDAR DATE	ROTATION NUMBER	HELIOGRAPHIC			DIAMETER	RA	DEC
		Lo	Bo	Po	(arcmin)	(HH:MM)	(deg)
8/16/2010	2100	275.56	6.68	16.27	31.588	09:41.9	13.85
8/17/2010	2100	262.34	6.72	16.60	31.594	09:45.6	13.53
8/18/2010	2100	249.12	6.77	16.93	31.600	09:49.3	13.21
8/19/2010	2100	235.90	6.81	17.26	31.606	09:53.1	12.89
8/20/2010	2100	222.68	6.85	17.58	31.613	09:56.8	12.56
8/21/2010	2100	209.46	6.89	17.89	31.619	10:0 .5	12.23
8/22/2010	2100	196.24	6.93	18.20	31.625	10: 4.2	11.90
8/23/2010	2100	183.03	6.96	18.51	31.632	10: 7.9	11.56
8/24/2010	2100	169.81	7.00	18.81	31.638	10:11.5	11.22
8/25/2010	2100	156.59	7.03	19.10	31.645	10:15.2	10.88
8/26/2010	2100	143.44	7.06	19.39	31.652	10:18.9	10.53
8/27/2010	2100	130.22	7.08	19.67	31.659	10:22.5	10.18
8/28/2010	2100	117.01	7.11	19.95	31.666	10:26.2	9.83
8/29/2010	2100	103.79	7.13	20.23	31.673	10:29.8	9.48
8/30/2010	2100	90.58	7.15	20.50	31.680	10:33.5	9.12
8/31/2010	2100	77.37	7.17	20.76	31.688	10:37.1	8.77
9/ 1/2010	2100	64.16	7.19	21.02	31.695	10:40.8	8.40
9/ 2/2010	2100	50.94	7.20	21.27	31.703	10:44.4	8.04
9/ 3/2010	2100	37.73	7.22	21.51	31.710	10:48.0	7.68
9/ 4/2010	2100	24.52	7.23	21.75	31.718	10:51.6	7.31
9/ 5/2010	2100	11.31	7.24	21.99	31.726	10:55.2	6.94
9/ 6/2010	2101	358.10	7.24	22.22	31.733	10:58.9	6.57
9/ 7/2010	2101	344.89	7.25	22.44	31.741	11: 2.5	6.20
9/ 8/2010	2101	331.68	7.25	22.66	31.749	11: 6.1	5.82
9/ 9/2010	2101	318.48	7.25	22.87	31.757	11: 9.7	5.44
9/10/2010	2101	305.27	7.25	23.08	31.766	11:13.3	5.07
9/11/2010	2101	292.06	7.24	23.28	31.774	11:16.9	4.69
9/12/2010	2101	278.85	7.24	23.47	31.782	11:20.4	4.31
9/13/2010	2101	265.65	7.23	23.66	31.790	11:24.0	3.92
9/14/2010	2101	252.44	7.22	23.84	31.799	11:27.6	3.54
9/15/2010	2101	239.24	7.21	24.02	31.807	11:31.2	3.16
9/16/2010	2101	226.10	7.19	24.19	31.816	11:34.8	2.77
9/17/2010	2101	212.89	7.18	24.35	31.825	11:38.4	2.39
9/18/2010	2101	199.69	7.16	24.51	31.833	11:42.0	2.00
9/19/2010	2101	186.48	7.14	24.66	31.842	11:45.6	1.61
9/20/2010	2101	173.28	7.12	24.80	31.851	11:49.2	1.22
9/21/2010	2101	160.08	7.09	24.94	31.859	11:52.7	0.83
9/22/2010	2101	146.88	7.06	25.07	31.868	11:56.3	0.45
9/23/2010	2101	133.68	7.04	25.20	31.877	11:59.9	0.06
9/24/2010	2101	120.47	7.00	25.32	31.886	12: 3.5	−0.33
9/25/2010	2101	107.27	6.97	25.43	31.895	12: 7.1	−0.72
9/26/2010	2101	94.07	6.94	25.53	31.904	12:10.7	−1.11
9/27/2010	2101	80.87	6.90	25.63	31.913	12:14.3	−1.50
9/28/2010	2101	67.67	6.86	25.72	31.922	12:17.9	−1.89
9/29/2010	2101	54.48	6.82	25.81	31.931	12:21.5	−2.28
9/30/2010	2101	41.28	6.77	25.89	31.940	12:25.2	−2.67

CALENDAR DATE	ROTATION NUMBER	HELIOGRAPHIC			DIAMETER	RA	DEC
		Lo	Bo	Po	(arcmin)	(HH:MM)	(deg)
10/ 1/2010	2101	28.08	6.73	25.96	31.949	12:28.8	−3.06
10/ 2/2010	2101	14.88	6.68	26.02	31.959	12:32.4	−3.45
10/ 3/2010	2101	1.68	6.63	26.08	31.968	12:36.0	−3.84
10/ 4/2010	2102	348.49	6.58	26.13	31.977	12:39.7	−4.22
10/ 5/2010	2102	335.29	6.53	26.17	31.986	12:43.3	−4.61
10/ 6/2010	2102	322.16	6.47	26.21	31.995	12:46.9	−4.99
10/ 7/2010	2102	308.96	6.41	26.24	32.005	12:50.6	−5.38
10/ 8/2010	2102	295.76	6.36	26.26	32.014	12:54.3	−5.76
10/ 9/2010	2102	282.57	6.29	26.27	32.023	12:57.9	−6.14
10/10/2010	2102	269.37	6.23	26.28	32.032	13: 1.6	−6.52
10/11/2010	2102	256.18	6.17	26.28	32.041	13: 5.3	−6.90
10/12/2010	2102	242.99	6.10	26.27	32.051	13: 9.0	−7.28
10/13/2010	2102	229.79	6.03	26.26	32.060	13:12.7	−7.65
10/14/2010	2102	216.60	5.96	26.23	32.069	13:16.4	−8.03
10/15/2010	2102	203.40	5.89	26.20	32.078	13:20.1	−8.40
10/16/2010	2102	190.21	5.81	26.17	32.087	13:23.8	−8.77
10/17/2010	2102	177.02	5.74	26.12	32.096	13:27.5	−9.14
10/18/2010	2102	163.83	5.66	26.07	32.106	13:31.3	−9.50
10/19/2010	2102	150.63	5.58	26.01	32.115	13:35.0	−9.87
10/20/2010	2102	137.44	5.50	25.94	32.124	13:38.8	−10.23
10/21/2010	2102	124.25	5.42	25.86	32.133	13:42.6	−10.59
10/22/2010	2102	111.06	5.33	25.78	32.142	13:46.4	−10.94
10/23/2010	2102	97.87	5.25	25.69	32.151	13:50.2	−11.29
10/24/2010	2102	84.68	5.16	25.59	32.159	13:54.0	−11.64
10/25/2010	2102	71.49	5.07	25.48	32.168	13:57.8	−11.99
10/26/2010	2102	58.36	4.98	25.37	32.177	14: 1.6	−12.34
10/27/2010	2102	45.17	4.88	25.25	32.186	14: 5.5	−12.68
10/28/2010	2102	31.98	4.79	25.12	32.194	14: 9.3	−13.02
10/29/2010	2102	18.79	4.70	24.98	32.203	14:13.2	−13.35
10/30/2010	2102	5.60	4.60	24.83	32.212	14:17.1	−13.68
10/31/2010	2103	352.41	4.50	24.68	32.220	14:20.9	−14.01
11/ 1/2010	2103	339.22	4.40	24.52	32.229	14:24.9	−14.33
11/ 2/2010	2103	326.03	4.30	24.35	32.237	14:28.8	−14.65
11/ 3/2010	2103	312.84	4.20	24.17	32.245	14:32.7	−14.97
11/ 4/2010	2103	299.66	4.09	23.99	32.254	14:36.7	−15.28
11/ 5/2010	2103	286.47	3.99	23.79	32.262	14:40.6	−15.59
11/ 6/2010	2103	273.28	3.88	23.59	32.270	14:44.6	−15.89
11/ 7/2010	2103	260.09	3.77	23.38	32.278	14:48.6	−16.19
11/ 8/2010	2103	246.90	3.66	23.17	32.286	14:52.6	−16.48
11/ 9/2010	2103	233.72	3.55	22.94	32.294	14:56.6	−16.77
11/10/2010	2103	220.53	3.44	22.71	32.301	15:0 .6	−17.06
11/11/2010	2103	207.34	3.33	22.47	32.309	15: 4.7	−17.34
11/12/2010	2103	194.16	3.22	22.22	32.317	15: 8.7	−17.61
11/13/2010	2103	180.97	3.10	21.97	32.324	15:12.8	−17.88
11/14/2010	2103	167.78	2.99	21.71	32.331	15:16.9	−18.15
11/15/2010	2103	154.66	2.87	21.43	32.339	15:21.0	−18.41

CALENDAR DATE	ROTATION NUMBER	HELIOGRAPHIC			DIAMETER	RA	DEC
		Lo	Bo	Po	(arcmin)	(HH:MM)	(deg)
11/16/2010	2103	141.48	2.75	21.16	32.346	15:25.1	−18.66
11/17/2010	2103	128.29	2.64	20.87	32.353	15:29.2	−18.91
11/18/2010	2103	115.10	2.52	20.58	32.360	15:33.4	−19.15
11/19/2010	2103	101.92	2.40	20.28	32.367	15:37.5	−19.39
11/20/2010	2103	88.74	2.28	19.97	32.374	15:41.7	−19.62
11/21/2010	2103	75.55	2.15	19.66	32.380	15:45.9	−19.85
11/22/2010	2103	62.37	2.03	19.34	32.387	15:50.1	−20.07
11/23/2010	2103	49.18	1.91	19.01	32.393	15:54.3	−20.28
11/24/2010	2103	36.00	1.79	18.67	32.399	15:58.5	−20.49
11/25/2010	2103	22.82	1.66	18.33	32.405	16: 2.7	−20.69
11/26/2010	2103	9.63	1.54	17.98	32.411	16: 7.0	−20.88
11/27/2010	2104	356.45	1.41	17.63	32.417	16:11.2	−21.07
11/28/2010	2104	343.27	1.29	17.27	32.423	16:15.5	−21.25
11/29/2010	2104	330.08	1.16	16.90	32.429	16:19.8	−21.43
11/30/2010	2104	316.90	1.03	16.52	32.434	16:24.1	−21.60
12/ 1/2010	2104	303.72	0.91	16.14	32.440	16:28.4	−21.76
12/ 2/2010	2104	290.54	0.78	15.76	32.445	16:32.7	−21.91
12/ 3/2010	2104	277.35	0.65	15.36	32.450	16:37.0	−22.06
12/ 4/2010	2104	264.17	0.53	14.97	32.455	16:41.4	−22.20
12/ 5/2010	2104	251.05	0.40	14.56	32.460	16:45.7	−22.33
12/ 6/2010	2104	237.87	0.27	14.15	32.464	16:50.1	−22.46
12/ 7/2010	2104	224.69	0.14	13.74	32.469	16:54.4	−22.57
12/ 8/2010	2104	211.51	0.01	13.32	32.473	16:58.8	−22.69
12/ 9/2010	2104	198.33	−0.11	12.90	32.477	17: 3.2	−22.79
12/10/2010	2104	185.15	−0.24	12.47	32.482	17: 7.6	−22.89
12/11/2010	2104	171.97	−0.37	12.03	32.485	17:12.0	−22.97
12/12/2010	2104	158.79	−0.50	11.59	32.489	17:16.4	−23.06
12/13/2010	2104	145.61	−0.63	11.15	32.493	17:20.8	−23.13
12/14/2010	2104	132.43	−0.75	10.71	32.496	17:25.2	−23.19
12/15/2010	2104	119.25	−0.88	10.26	32.500	17:29.6	−23.25
12/16/2010	2104	106.08	−1.01	9.80	32.503	17:34.0	−23.30
12/17/2010	2104	92.90	−1.14	9.35	32.506	17:38.5	−23.35
12/18/2010	2104	79.72	−1.26	8.89	32.509	17:42.9	−23.38
12/19/2010	2104	66.54	−1.39	8.42	32.511	17:47.3	−23.41
12/20/2010	2104	53.36	−1.51	7.96	32.514	17:51.7	−23.43
12/21/2010	2104	40.19	−1.64	7.49	32.516	17:56.2	−23.44
12/22/2010	2104	27.01	−1.76	7.02	32.518	18:0.6	−23.44
12/23/2010	2104	13.83	−1.89	6.54	32.520	18: 5.1	−23.44
12/24/2010	2104	0.66	−2.01	6.07	32.522	18: 9.5	−23.43
12/25/2010	2105	347.54	−2.14	5.59	32.524	18:13.9	−23.41
12/26/2010	2105	334.37	−2.26	5.11	32.525	18:18.4	−23.38
12/27/2010	2105	321.19	−2.38	4.63	32.527	18:22.8	−23.34
12/28/2010	2105	308.02	−2.50	4.15	32.528	18:27.2	−23.30
12/29/2010	2105	294.84	−2.62	3.67	32.529	18:31.7	−23.25
12/30/2010	2105	281.67	−2.74	3.18	32.530	18:36.1	−23.19
12/31/2010	2105	268.49	−2.86	2.70	32.530	18:40.5	−23.12

CALENDAR DATE	ROTATION NUMBER	HELIOGRAPHIC			DIAMETER	RA	DEC
		Lo	Bo	Po	(arcmin)	(HH:MM)	(deg)
1/1/2011	2105	255.32	−2.98	2.21	32.531	18:44.9	−23.05
1/2/2011	2105	242.14	−3.09	1.73	32.531	18:49.3	−22.97
1/3/2011	2105	228.97	−3.21	1.24	32.531	18:53.7	−22.88
1/4/2011	2105	215.79	−3.32	0.76	32.531	18:58.1	−22.78
1/5/2011	2105	202.62	−3.44	0.28	32.531	19: 2.5	−22.67
1/6/2011	2105	189.45	−3.55	−0.21	32.531	19: 6.9	−22.56
1/7/2011	2105	176.27	−3.66	−0.69	32.530	19:11.3	−22.44
1/8/2011	2105	163.10	−3.77	−1.17	32.530	19:15.7	−22.32
1/9/2011	2105	149.93	−3.88	−1.65	32.529	19:20.0	−22.18
1/10/2011	2105	136.76	−3.99	−2.13	32.528	19:24.4	−22.04
1/11/2011	2105	123.58	−4.10	−2.61	32.527	19:28.7	−21.89
1/12/2011	2105	110.41	−4.20	−3.09	32.525	19:33.1	−21.74
1/13/2011	2105	97.24	−4.31	−3.56	32.524	19:37.4	−21.57
1/14/2011	2105	84.13	−4.41	−4.03	32.522	19:41.7	−21.40
1/15/2011	2105	70.96	−4.51	−4.50	32.520	19:46.0	−21.23
1/16/2011	2105	57.79	−4.61	−4.97	32.518	19:50.3	−21.04
1/17/2011	2105	44.62	−4.71	−5.44	32.516	19:54.6	−20.85
1/18/2011	2105	31.45	−4.81	−5.90	32.514	19:58.9	−20.66
1/19/2011	2105	18.28	−4.90	−6.36	32.511	20: 3.1	−20.45
1/20/2011	2105	5.10	−5.00	−6.82	32.509	20: 7.4	−20.24
1/21/2011	2106	351.93	−5.09	−7.27	32.506	20:11.6	−20.03
1/22/2011	2106	338.76	−5.18	−7.72	32.503	20:15.8	−19.81
1/23/2011	2106	325.59	−5.27	−8.17	32.500	20:20.0	−19.58
1/24/2011	2106	312.42	−5.36	−8.61	32.496	20:24.2	−19.34
1/25/2011	2106	299.25	−5.44	−9.05	32.493	20:28.4	−19.10
1/26/2011	2106	286.08	−5.53	−9.49	32.489	20:32.6	−18.86
1/27/2011	2106	272.91	−5.61	−9.92	32.485	20:36.8	−18.60
1/28/2011	2106	259.74	−5.69	−10.35	32.482	20:40.9	−18.35
1/29/2011	2106	246.57	−5.77	−10.78	32.477	20:45.0	−18.08
1/30/2011	2106	233.40	−5.85	−11.20	32.473	20:49.1	−17.81
1/31/2011	2106	220.23	−5.92	−11.61	32.469	20:53.2	−17.54
2/1/2011	2106	207.06	−5.99	−12.02	32.464	20:57.3	−17.26
2/2/2011	2106	193.89	−6.07	−12.43	32.460	21: 1.4	−16.98
2/3/2011	2106	180.72	−6.13	−12.83	32.455	21: 5.5	−16.69
2/4/2011	2106	167.62	−6.20	−13.23	32.450	21: 9.5	−16.40
2/5/2011	2106	154.45	−6.27	−13.62	32.445	21:13.6	−16.10
2/6/2011	2106	141.28	−6.33	−14.01	32.440	21:17.6	−15.79
2/7/2011	2106	128.11	−6.39	−14.39	32.434	21:21.6	−15.49
2/8/2011	2106	114.94	−6.45	−14.77	32.429	21:25.6	−15.17
2/9/2011	2106	101.77	−6.51	−15.14	32.423	21:29.6	−14.86
2/10/2011	2106	88.59	−6.57	−15.51	32.417	21:33.5	−14.54
2/11/2011	2106	75.42	−6.62	−15.87	32.411	21:37.5	−14.21
2/12/2011	2106	62.25	−6.67	−16.23	32.405	21:41.4	−13.88
2/13/2011	2106	49.08	−6.72	−16.58	32.399	21:45.4	−13.55
2/14/2011	2106	35.91	−6.77	−16.93	32.393	21:49.3	−13.21
2/15/2011	2106	22.74	−6.81	−17.27	32.387	21:53.2	−12.87

CALENDAR DATE	ROTATION NUMBER	HELIOGRAPHIC			DIAMETER	RA	DEC
		Lo	Bo	Po	(arcmin)	(HH:MM)	(deg)
2/16/2011	2106	9.57	−6.85	−17.60	32.380	21:57.1	−12.53
2/17/2011	2107	356.40	−6.89	−17.93	32.374	22: 1.0	−12.18
2/18/2011	2107	343.23	−6.93	−18.26	32.367	22: 4.9	−11.84
2/19/2011	2107	330.06	−6.97	−18.58	32.360	22: 8.7	−11.48
2/20/2011	2107	316.88	−7.00	−18.89	32.353	22:12.6	−11.13
2/21/2011	2107	303.71	−7.04	−19.19	32.346	22:16.4	−10.77
2/22/2011	2107	290.54	−7.07	−19.49	32.339	22:20.2	−10.41
2/23/2011	2107	277.36	−7.09	−19.79	32.331	22:24.0	−10.04
2/24/2011	2107	264.25	−7.12	−20.08	32.324	22:27.8	−9.67
2/25/2011	2107	251.08	−7.14	−20.36	32.317	22:31.6	−9.31
2/26/2011	2107	237.91	−7.16	−20.64	32.309	22:35.4	−8.93
2/27/2011	2107	224.73	−7.18	−20.91	32.301	22:39.2	−8.56
2/28/2011	2107	211.56	−7.20	−21.17	32.294	22:43.0	−8.18
3/ 1/2011	2107	198.38	−7.21	−21.43	32.286	22:46.7	−7.81
3/ 2/2011	2107	185.21	−7.22	−21.68	32.278	22:50.5	−7.43
3/ 3/2011	2107	172.03	−7.23	−21.92	32.270	22:54.2	−7.04
3/ 4/2011	2107	158.85	−7.24	−22.16	32.262	22:58.0	−6.66
3/ 5/2011	2107	145.68	−7.25	−22.39	32.254	23: 1.7	−6.28
3/ 6/2011	2107	132.50	−7.25	−22.62	32.245	23: 5.4	−5.89
3/ 7/2011	2107	119.32	−7.25	−22.84	32.237	23: 9.1	−5.50
3/ 8/2011	2107	106.14	−7.25	−23.05	32.229	23:12.8	−5.11
3/ 9/2011	2107	92.96	−7.24	−23.26	32.220	23:16.5	−4.72
3/10/2011	2107	79.78	−7.24	−23.46	32.212	23:20.2	−4.33
3/11/2011	2107	66.60	−7.23	−23.65	32.203	23:23.9	−3.94
3/12/2011	2107	53.42	−7.22	−23.84	32.194	23:27.6	−3.55
3/13/2011	2107	40.24	−7.21	−24.02	32.186	23:31.3	−3.15
3/14/2011	2107	27.06	−7.19	−24.19	32.177	23:34.9	−2.76
3/15/2011	2107	13.88	−7.18	−24.36	32.168	23:38.6	−2.36
3/16/2011	2108	0.76	−7.16	−24.52	32.159	23:42.3	−1.97
3/17/2011	2108	347.57	−7.14	−24.67	32.151	23:45.9	−1.57
3/18/2011	2108	334.39	−7.11	−24.82	32.142	23:49.6	−1.18
3/19/2011	2108	321.21	−7.09	−24.96	32.133	23:53.2	−0.78
3/20/2011	2108	308.02	−7.06	−25.09	32.124	23:56.9	−0.39
3/21/2011	2108	294.83	−7.03	−25.22	32.115	00:0 .5	0.01
3/22/2011	2108	281.65	−7.00	−25.34	32.106	00: 4.2	0.40
3/23/2011	2108	268.46	−6.96	−25.45	32.096	00: 7.8	0.80
3/24/2011	2108	255.27	−6.93	−25.55	32.087	00:11.5	1.19
3/25/2011	2108	242.08	−6.89	−25.65	32.078	00:15.1	1.59
3/26/2011	2108	228.89	−6.85	−25.74	32.069	00:18.8	1.98
3/27/2011	2108	215.70	−6.81	−25.83	32.060	00:22.4	2.37
3/28/2011	2108	202.51	−6.76	−25.90	32.051	00:26.0	2.77
3/29/2011	2108	189.32	−6.72	−25.97	32.041	00:29.7	3.16
3/30/2011	2108	176.13	−6.67	−26.04	32.032	00:33.3	3.55
3/31/2011	2108	162.93	−6.62	−26.09	32.023	00:37.0	3.94

CALENDAR DATE	ROTATION NUMBER	HELIOGRAPHIC			DIAMETER	RA	DEC
		Lo	Bo	Po	(arcmin)	(HH:MM)	(deg)
4/ 1/2011	2108	149.74	−6.57	−26.14	32.014	00:40.6	4.32
4/ 2/2011	2108	136.54	−6.51	−26.18	32.005	00:44.3	4.71
4/ 3/2011	2108	123.35	−6.46	−26.21	31.995	00:47.9	5.09
4/ 4/2011	2108	110.15	−6.40	−26.24	31.986	00:51.6	5.48
4/ 5/2011	2108	97.02	−6.34	−26.26	31.977	00:55.2	5.86
4/ 6/2011	2108	83.82	−6.28	−26.27	31.968	00:58.9	6.24
4/ 7/2011	2108	70.62	−6.22	−26.28	31.959	01: 2.5	6.62
4/ 8/2011	2108	57.42	−6.15	−26.28	31.949	01: 6.2	6.99
4/ 9/2011	2108	44.22	−6.08	−26.27	31.940	01: 9.9	7.37
4/10/2011	2108	31.02	−6.01	−26.25	31.931	01:13.5	7.74
4/11/2011	2108	17.81	−5.94	−26.23	31.922	01:17.2	8.11
4/12/2011	2108	4.61	−5.87	−26.19	31.913	01:20.9	8.48
4/13/2011	2109	351.41	−5.80	−26.16	31.904	01:24.6	8.84
4/14/2011	2109	338.20	−5.72	−26.11	31.895	01:28.3	9.21
4/15/2011	2109	325.00	−5.65	−26.06	31.886	01:32.0	9.57
4/16/2011	2109	311.79	−5.57	−26.00	31.877	01:35.7	9.93
4/17/2011	2109	298.58	−5.49	−25.93	31.868	01:39.4	10.28
4/18/2011	2109	285.38	−5.41	−25.85	31.859	01:43.1	10.63
4/19/2011	2109	272.17	−5.32	−25.77	31.851	01:46.8	10.98
4/20/2011	2109	258.96	−5.24	−25.68	31.842	01:50.5	11.33
4/21/2011	2109	245.75	−5.15	−25.58	31.833	01:54.3	11.67
4/22/2011	2109	232.54	−5.06	−25.48	31.825	01:58.0	12.01
4/23/2011	2109	219.32	−4.97	−25.36	31.816	02: 1.7	12.35
4/24/2011	2109	206.11	−4.88	−25.24	31.807	02: 5.5	12.68
4/25/2011	2109	192.96	−4.79	−25.12	31.799	02: 9.3	13.01
4/26/2011	2109	179.75	−4.70	−24.98	31.790	02:13.0	13.34
4/27/2011	2109	166.53	−4.61	−24.84	31.782	02:16.8	13.66
4/28/2011	2109	153.31	−4.51	−24.69	31.774	02:20.6	13.98
4/29/2011	2109	140.10	−4.41	−24.54	31.766	02:24.4	14.29
4/30/2011	2109	126.88	−4.31	−24.37	31.757	02:28.2	14.61
5/ 1/2011	2109	113.66	−4.21	−24.20	31.749	02:32.0	14.91
5/ 2/2011	2109	100.44	−4.11	−24.02	31.741	02:35.8	15.21
5/ 3/2011	2109	87.22	−4.01	−23.84	31.733	02:39.7	15.51
5/ 4/2011	2109	74.00	−3.91	−23.65	31.726	02:43.5	15.81
5/ 5/2011	2109	60.78	−3.81	−23.45	31.718	02:47.4	16.10
5/ 6/2011	2109	47.56	−3.70	−23.24	31.710	02:51.2	16.38
5/ 7/2011	2109	34.33	−3.60	−23.03	31.703	02:55.1	16.66
5/ 8/2011	2109	21.11	−3.49	−22.81	31.695	02:59.0	16.94
5/ 9/2011	2109	7.89	−3.38	−22.58	31.688	03: 2.8	17.21
5/10/2011	2110	354.66	−3.27	−22.34	31.680	03: 6.7	17.48
5/11/2011	2110	341.44	−3.16	−22.10	31.673	03:10.6	17.74
5/12/2011	2110	328.21	−3.05	−21.85	31.666	03:14.6	18.00
5/13/2011	2110	314.98	−2.94	−21.60	31.659	03:18.5	18.25
5/14/2011	2110	301.76	−2.83	−21.34	31.652	03:22.4	18.50
5/15/2011	2110	288.59	−2.72	−21.07	31.645	03:26.4	18.74

CALENDAR DATE	ROTATION NUMBER	HELIOGRAPHIC			DIAMETER	RA	DEC
		Lo	Bo	Po	(arcmin)	(HH:MM)	(deg)
5/16/2011	2110	275.36	−2.60	−20.79	31.638	03:30.3	18.98
5/17/2011	2110	262.13	−2.49	−20.51	31.632	03:34.3	19.21
5/18/2011	2110	248.90	−2.38	−20.23	31.625	03:38.2	19.43
5/19/2011	2110	235.67	−2.26	−19.93	31.619	03:42.2	19.65
5/20/2011	2110	222.44	−2.14	−19.63	31.613	03:46.2	19.87
5/21/2011	2110	209.21	−2.03	−19.32	31.606	03:50.2	20.08
5/22/2011	2110	195.98	−1.91	−19.01	31.600	03:54.2	20.28
5/23/2011	2110	182.74	−1.79	−18.69	31.594	03:58.2	20.48
5/24/2011	2110	169.51	−1.68	−18.37	31.588	04: 2.3	20.67
5/25/2011	2110	156.28	−1.56	−18.04	31.583	04: 6.3	20.85
5/26/2011	2110	143.04	−1.44	−17.70	31.577	04:10.3	21.03
5/27/2011	2110	129.81	−1.32	−17.36	31.572	04:14.4	21.21
5/28/2011	2110	116.57	−1.20	−17.01	31.566	04:18.4	21.37
5/29/2011	2110	103.34	−1.08	−16.66	31.561	04:22.5	21.53
5/30/2011	2110	90.10	−0.96	−16.30	31.556	04:26.6	21.69
5/31/2011	2110	76.87	−0.84	−15.94	31.551	04:30.7	21.84
6/ 1/2011	2110	63.63	−0.72	−15.57	31.546	04:34.7	21.98
6/ 2/2011	2110	50.39	−0.60	−15.20	31.541	04:38.8	22.12
6/ 3/2011	2110	37.15	−0.48	−14.82	31.537	04:42.9	22.25
6/ 4/2011	2110	23.98	−0.36	−14.44	31.532	04:47.0	22.37
6/ 5/2011	2110	10.74	−0.24	−14.05	31.528	04:51.1	22.49
6/ 6/2011	2111	357.50	−0.12	−13.66	31.524	04:55.3	22.60
6/ 7/2011	2111	344.26	0.00	−13.26	31.519	04:59.4	22.70
6/ 8/2011	2111	331.03	0.12	−12.86	31.515	05: 3.5	22.80
6/ 9/2011	2111	317.79	0.24	−12.46	31.512	05: 7.6	22.89
6/10/2011	2111	304.55	0.36	−12.05	31.508	05:11.8	22.97
6/11/2011	2111	291.31	0.48	−11.64	31.504	05:15.9	23.05
6/12/2011	2111	278.07	0.60	−11.22	31.501	05:20.1	23.12
6/13/2011	2111	264.83	0.72	−10.81	31.498	05:24.2	23.18
6/14/2011	2111	251.59	0.84	−10.38	31.495	05:28.4	23.24
6/15/2011	2111	238.35	0.96	−9.96	31.492	05:32.5	23.29
6/16/2011	2111	225.11	1.08	−9.53	31.489	05:36.7	23.33
6/17/2011	2111	211.87	1.20	−9.10	31.486	05:40.8	23.37
6/18/2011	2111	198.63	1.32	−8.67	31.483	05:45.0	23.39
6/19/2011	2111	185.39	1.44	−8.23	31.481	05:49.1	23.42
6/20/2011	2111	172.15	1.56	−7.79	31.479	05:53.3	23.43
6/21/2011	2111	158.91	1.67	−7.35	31.477	05:57.4	23.44
6/22/2011	2111	145.66	1.79	−6.91	31.475	06: 1.6	23.44
6/23/2011	2111	132.42	1.91	−6.47	31.473	06: 5.8	23.44
6/24/2011	2111	119.18	2.02	−6.02	31.471	06: 9.9	23.42
6/25/2011	2111	106.00	2.14	−5.58	31.469	06:14.1	23.40
6/26/2011	2111	92.76	2.25	−5.13	31.468	06:18.2	23.38
6/27/2011	2111	79.52	2.37	−4.68	31.467	06:22.4	23.35
6/28/2011	2111	66.28	2.48	−4.23	31.466	06:26.5	23.31
6/29/2011	2111	53.04	2.59	−3.78	31.465	06:30.7	23.26
6/30/2011	2111	39.80	2.71	−3.32	31.464	06:34.8	23.21

CALENDAR DATE	ROTATION NUMBER	HELIOGRAPHIC			DIAMETER	RA	DEC
		Lo	Bo	Po	(arcmin)	(HH:MM)	(deg)
7/ 1/2011	2111	26.56	2.82	−2.87	31.463	06:38.9	23.15
7/ 2/2011	2111	13.32	2.93	−2.42	31.463	06:43.1	23.08
7/ 3/2011	2111	0.08	3.04	−1.96	31.462	06:47.2	23.01
7/ 4/2011	2112	346.84	3.15	−1.51	31.462	06:51.3	22.93
7/ 5/2011	2112	333.60	3.25	−1.06	31.462	06:55.4	22.84
7/ 6/2011	2112	320.36	3.36	−0.60	31.462	06:59.6	22.75
7/ 7/2011	2112	307.12	3.47	−0.15	31.462	07: 3.7	22.65
7/ 8/2011	2112	293.89	3.57	0.30	31.462	07: 7.8	22.54
7/ 9/2011	2112	280.65	3.68	0.75	31.463	07:11.9	22.43
7/10/2011	2112	267.41	3.78	1.20	31.464	07:16.0	22.31
7/11/2011	2112	254.17	3.88	1.65	31.464	07:20.0	22.18
7/12/2011	2112	240.93	3.98	2.10	31.465	07:24.1	22.05
7/13/2011	2112	227.70	4.08	2.55	31.466	07:28.2	21.91
7/14/2011	2112	214.46	4.18	3.00	31.468	07:32.2	21.77
7/15/2011	2112	201.28	4.28	3.44	31.469	07:36.3	21.62
7/16/2011	2112	188.05	4.38	3.88	31.471	07:40.3	21.46
7/17/2011	2112	174.81	4.47	4.32	31.472	07:44.4	21.30
7/18/2011	2112	161.58	4.57	4.76	31.474	07:48.4	21.13
7/19/2011	2112	148.34	4.66	5.20	31.476	07:52.4	20.95
7/20/2011	2112	135.11	4.75	5.63	31.478	07:56.4	20.77
7/21/2011	2112	121.87	4.84	6.07	31.480	07:60.4	20.58
7/22/2011	2112	108.64	4.93	6.50	31.483	08: 4.4	20.39
7/23/2011	2112	95.40	5.02	6.93	31.485	08: 8.4	20.19
7/24/2011	2112	82.17	5.11	7.35	31.488	08:12.4	19.99
7/25/2011	2112	68.94	5.19	7.77	31.491	08:16.3	19.78
7/26/2011	2112	55.70	5.27	8.19	31.494	08:20.3	19.56
7/27/2011	2112	42.47	5.36	8.61	31.497	08:24.2	19.34
7/28/2011	2112	29.24	5.44	9.02	31.500	08:28.1	19.12
7/29/2011	2112	16.01	5.52	9.43	31.504	08:32.1	18.89
7/30/2011	2112	2.78	5.59	9.84	31.507	08:36.0	18.65
7/31/2011	2113	349.55	5.67	10.25	31.511	08:39.9	18.41
8/ 1/2011	2113	336.32	5.74	10.65	31.515	08:43.8	18.16
8/ 2/2011	2113	323.09	5.82	11.04	31.518	08:47.7	17.91
8/ 3/2011	2113	309.86	5.89	11.44	31.523	08:51.5	17.66
8/ 4/2011	2113	296.69	5.96	11.83	31.527	08:55.4	17.40
8/ 5/2011	2113	283.47	6.03	12.21	31.531	08:59.2	17.13
8/ 6/2011	2113	270.24	6.09	12.59	31.535	09: 3.1	16.86
8/ 7/2011	2113	257.01	6.16	12.97	31.540	09: 6.9	16.59
8/ 8/2011	2113	243.79	6.22	13.35	31.545	09:10.7	16.31
8/ 9/2011	2113	230.56	6.28	13.72	31.550	09:14.5	16.02
8/10/2011	2113	217.34	6.34	14.08	31.555	09:18.4	15.74
8/11/2011	2113	204.11	6.40	14.44	31.560	09:22.1	15.44
8/12/2011	2113	190.89	6.46	14.80	31.565	09:25.9	15.15
8/13/2011	2113	177.67	6.51	15.15	31.570	09:29.7	14.85
8/14/2011	2113	164.44	6.56	15.50	31.576	09:33.5	14.54
8/15/2011	2113	151.22	6.61	15.85	31.581	09:37.2	14.23

CALENDAR DATE	ROTATION NUMBER	HELIOGRAPHIC			DIAMETER	RA	DEC
		Lo	Bo	Po	(arcmin)	(HH:MM)	(deg)
8/16/2011	2113	138.00	6.66	16.19	31.587	09:41.0	13.92
8/17/2011	2113	124.78	6.71	16.52	31.593	09:44.7	13.61
8/18/2011	2113	111.56	6.76	16.85	31.599	09:48.4	13.29
8/19/2011	2113	98.34	6.80	17.18	31.605	09:52.2	12.97
8/20/2011	2113	85.12	6.84	17.50	31.611	09:55.9	12.64
8/21/2011	2113	71.90	6.88	17.81	31.617	09:59.6	12.31
8/22/2011	2113	58.68	6.92	18.13	31.624	10: 3.3	11.98
8/23/2011	2113	45.46	6.95	18.43	31.630	10: 7.0	11.64
8/24/2011	2113	32.31	6.99	18.73	31.637	10:10.7	11.30
8/25/2011	2113	19.09	7.02	19.03	31.644	10:14.3	10.96
8/26/2011	2113	5.88	7.05	19.32	31.650	10:18.0	10.62
8/27/2011	2114	352.66	7.08	19.60	31.657	10:21.7	10.27
8/28/2011	2114	339.45	7.10	19.88	31.664	10:25.3	9.92
8/29/2011	2114	326.23	7.13	20.16	31.671	10:29.0	9.57
8/30/2011	2114	313.02	7.15	20.43	31.679	10:32.6	9.21
8/31/2011	2114	299.81	7.17	20.69	31.686	10:36.2	8.85
9/ 1/2011	2114	286.59	7.18	20.95	31.693	10:39.9	8.49
9/ 2/2011	2114	273.38	7.20	21.21	31.701	10:43.5	8.13
9/ 3/2011	2114	260.17	7.21	21.45	31.708	10:47.1	7.77
9/ 4/2011	2114	246.96	7.22	21.69	31.716	10:50.8	7.40
9/ 5/2011	2114	233.75	7.23	21.93	31.724	10:54.4	7.03
9/ 6/2011	2114	220.54	7.24	22.16	31.732	10:58.0	6.66
9/ 7/2011	2114	207.33	7.25	22.39	31.739	11: 1.6	6.29
9/ 8/2011	2114	194.12	7.25	22.61	31.747	11: 5.2	5.91
9/ 9/2011	2114	180.91	7.25	22.82	31.755	11: 8.8	5.54
9/10/2011	2114	167.70	7.25	23.03	31.764	11:12.4	5.16
9/11/2011	2114	154.50	7.25	23.23	31.772	11:16.0	4.78
9/12/2011	2114	141.29	7.24	23.42	31.780	11:19.6	4.40
9/13/2011	2114	128.15	7.23	23.61	31.788	11:23.2	4.02
9/14/2011	2114	114.94	7.22	23.80	31.797	11:26.8	3.63
9/15/2011	2114	101.74	7.21	23.97	31.805	11:30.3	3.25
9/16/2011	2114	88.53	7.20	24.15	31.814	11:33.9	2.87
9/17/2011	2114	75.33	7.18	24.31	31.822	11:37.5	2.48
9/18/2011	2114	62.12	7.16	24.47	31.831	11:41.1	2.09
9/19/2011	2114	48.92	7.14	24.62	31.840	11:44.7	1.71
9/20/2011	2114	35.72	7.12	24.77	31.848	11:48.3	1.32
9/21/2011	2114	22.51	7.10	24.91	31.857	11:51.9	0.93
9/22/2011	2114	9.31	7.07	25.04	31.866	11:55.5	0.54
9/23/2011	2115	356.11	7.04	25.17	31.875	11:59.1	0.15
9/24/2011	2115	342.91	7.01	25.29	31.884	12: 2.7	−0.24
9/25/2011	2115	329.71	6.98	25.40	31.893	12: 6.3	−0.63
9/26/2011	2115	316.51	6.95	25.51	31.902	12: 9.9	−1.02
9/27/2011	2115	303.31	6.91	25.61	31.911	12:13.5	−1.41
9/28/2011	2115	290.11	6.87	25.70	31.920	12:17.1	−1.80
9/29/2011	2115	276.91	6.83	25.79	31.929	12:20.7	−2.19
9/30/2011	2115	263.71	6.79	25.87	31.938	12:24.3	−2.58

Appendix C

CALENDAR DATE	ROTATION NUMBER	HELIOGRAPHIC			DIAMETER	RA	DEC
		Lo	Bo	Po	(arcmin)	(HH:MM)	(deg)
10/ 1/2011	2115	250.51	6.74	25.94	31.947	12:27.9	–2.97
10/ 2/2011	2115	237.32	6.69	26.00	31.956	12:31.5	–3.35
10/ 3/2011	2115	224.18	6.65	26.06	31.966	12:35.2	–3.74
10/ 4/2011	2115	210.98	6.59	26.12	31.975	12:38.8	–4.13
10/ 5/2011	2115	197.79	6.54	26.16	31.984	12:42.4	–4.52
10/ 6/2011	2115	184.59	6.49	26.20	31.993	12:46.1	–4.90
10/ 7/2011	2115	171.39	6.43	26.23	32.002	12:49.7	–5.28
10/ 8/2011	2115	158.20	6.37	26.25	32.012	12:53.4	–5.67
10/ 9/2011	2115	145.00	6.31	26.27	32.021	12:57.0	–6.05
10/10/2011	2115	131.81	6.25	26.28	32.030	13:0 .7	–6.43
10/11/2011	2115	118.61	6.18	26.28	32.039	13: 4.4	–6.81
10/12/2011	2115	105.42	6.12	26.27	32.048	13: 8.1	–7.19
10/13/2011	2115	92.22	6.05	26.26	32.058	13:11.8	–7.56
10/14/2011	2115	79.03	5.98	26.24	32.067	13:15.5	–7.94
10/15/2011	2115	65.84	5.91	26.21	32.076	13:19.2	–8.31
10/16/2011	2115	52.64	5.83	26.17	32.085	13:22.9	–8.68
10/17/2011	2115	39.45	5.76	26.13	32.094	13:26.6	–9.05
10/18/2011	2115	26.26	5.68	26.08	32.103	13:30.4	–9.41
10/19/2011	2115	13.07	5.60	26.02	32.112	13:34.1	–9.78
10/20/2011	2115	359.87	5.52	25.96	32.121	13:37.9	–10.14
10/21/2011	2116	346.68	5.44	25.88	32.130	13:41.7	–10.50
10/22/2011	2116	333.49	5.35	25.80	32.139	13:45.4	–10.86
10/23/2011	2116	320.36	5.27	25.71	32.148	13:49.2	–11.21
10/24/2011	2116	307.17	5.18	25.61	32.157	13:53.0	–11.56
10/25/2011	2116	293.98	5.09	25.51	32.166	13:56.9	–11.91
10/26/2011	2116	280.79	5.00	25.40	32.175	14:0 .7	–12.25
10/27/2011	2116	267.60	4.91	25.28	32.184	14: 4.5	–12.60
10/28/2011	2116	254.41	4.82	25.15	32.192	14: 8.4	–12.93
10/29/2011	2116	241.22	4.72	25.01	32.201	14:12.2	–13.27
10/30/2011	2116	228.03	4.62	24.87	32.210	14:16.1	–13.60
10/31/2011	2116	214.84	4.53	24.72	32.218	14:20.0	–13.93
11/ 1/2011	2116	201.65	4.43	24.56	32.227	14:23.9	–14.25
11/ 2/2011	2116	188.46	4.32	24.39	32.235	14:27.8	–14.57
11/ 3/2011	2116	175.27	4.22	24.21	32.243	14:31.8	–14.89
11/ 4/2011	2116	162.09	4.12	24.03	32.252	14:35.7	–15.20
11/ 5/2011	2116	148.90	4.01	23.84	32.260	14:39.7	–15.51
11/ 6/2011	2116	135.71	3.91	23.64	32.268	14:43.6	–15.82
11/ 7/2011	2116	122.52	3.80	23.43	32.276	14:47.6	–16.12
11/ 8/2011	2116	109.34	3.69	23.22	32.284	14:51.6	–16.41
11/ 9/2011	2116	96.15	3.58	23.00	32.292	14:55.6	–16.70
11/10/2011	2116	82.96	3.47	22.77	32.299	14:59.6	–16.99
11/11/2011	2116	69.77	3.36	22.53	32.307	15: 3.7	–17.27
11/12/2011	2116	56.59	3.25	22.28	32.315	15: 7.7	–17.55
11/13/2011	2116	43.46	3.13	22.03	32.322	15:11.8	–17.82
11/14/2011	2116	30.28	3.02	21.77	32.330	15:15.9	–18.08
11/15/2011	2116	17.09	2.90	21.50	32.337	15:20.0	–18.35

CALENDAR DATE	ROTATION NUMBER	HELIOGRAPHIC			DIAMETER	RA	DEC
		Lo	Bo	Po	(arcmin)	(HH:MM)	(deg)
11/16/2011	2116	3.91	2.78	21.22	32.344	15:24.1	−18.60
11/17/2011	2117	350.72	2.67	20.94	32.351	15:28.2	−18.85
11/18/2011	2117	337.54	2.55	20.65	32.358	15:32.4	−19.09
11/19/2011	2117	324.35	2.43	20.35	32.365	15:36.5	−19.33
11/20/2011	2117	311.17	2.31	20.05	32.372	15:40.7	−19.57
11/21/2011	2117	297.98	2.19	19.73	32.379	15:44.8	−19.79
11/22/2011	2117	284.80	2.06	19.41	32.385	15:49.0	−20.01
11/23/2011	2117	271.61	1.94	19.09	32.391	15:53.2	−20.23
11/24/2011	2117	258.43	1.82	18.75	32.398	15:57.5	−20.44
11/25/2011	2117	245.25	1.69	18.41	32.404	16: 1.7	−20.64
11/26/2011	2117	232.06	1.57	18.07	32.410	16: 5.9	−20.84
11/27/2011	2117	218.88	1.44	17.71	32.416	16:10.2	−21.03
11/28/2011	2117	205.70	1.32	17.35	32.422	16:14.5	−21.21
11/29/2011	2117	192.51	1.19	16.99	32.427	16:18.7	−21.39
11/30/2011	2117	179.33	1.07	16.61	32.433	16:23.0	−21.56
12/ 1/2011	2117	166.15	0.94	16.23	32.438	16:27.3	−21.72
12/ 2/2011	2117	152.97	0.81	15.85	32.444	16:31.6	−21.87
12/ 3/2011	2117	139.85	0.69	15.46	32.449	16:36.0	−22.02
12/ 4/2011	2117	126.67	0.56	15.06	32.454	16:40.3	−22.16
12/ 5/2011	2117	113.48	0.43	14.66	32.459	16:44.7	−22.30
12/ 6/2011	2117	100.30	0.30	14.25	32.463	16:49.0	−22.43
12/ 7/2011	2117	87.12	0.18	13.84	32.468	16:53.4	−22.55
12/ 8/2011	2117	73.94	0.05	13.42	32.472	16:57.7	−22.66
12/ 9/2011	2117	60.76	−0.08	13.00	32.476	17: 2.1	−22.76
12/10/2011	2117	47.58	−0.21	12.57	32.481	17: 6.5	−22.86
12/11/2011	2117	34.40	−0.34	12.14	32.485	17:10.9	−22.95
12/12/2011	2117	21.22	−0.47	11.70	32.488	17:15.3	−23.04
12/13/2011	2117	8.04	−0.59	11.26	32.492	17:19.7	−23.11
12/14/2011	2118	354.86	−0.72	10.82	32.495	17:24.1	−23.18
12/15/2011	2118	341.68	−0.85	10.37	32.499	17:28.5	−23.24
12/16/2011	2118	328.51	−0.98	9.91	32.502	17:33.0	−23.29
12/17/2011	2118	315.33	−1.10	9.46	32.505	17:37.4	−23.34
12/18/2011	2118	302.15	−1.23	9.00	32.508	17:41.8	−23.37
12/19/2011	2118	288.97	−1.36	8.54	32.511	17:46.2	−23.40
12/20/2011	2118	275.79	−1.48	8.07	32.513	17:50.7	−23.42
12/21/2011	2118	262.62	−1.61	7.60	32.516	17:55.1	−23.44
12/22/2011	2118	249.44	−1.73	7.13	32.518	17:59.5	−23.44
12/23/2011	2118	236.32	−1.86	6.66	32.520	18: 4.0	−23.44
12/24/2011	2118	223.15	−1.98	6.18	32.522	18: 8.4	−23.43
12/25/2011	2118	209.97	−2.10	5.71	32.523	18:12.9	−23.41
12/26/2011	2118	196.80	−2.23	5.23	32.525	18:17.3	−23.39
12/27/2011	2118	183.62	−2.35	4.75	32.526	18:21.7	−23.35
12/28/2011	2118	170.44	−2.47	4.27	32.528	18:26.2	−23.31
12/29/2011	2118	157.27	−2.59	3.78	32.529	18:30.6	−23.26
12/30/2011	2118	144.09	−2.71	3.30	32.529	18:35.0	−23.20
12/31/2011	2118	130.92	−2.83	2.82	32.530	18:39.4	−23.14

CALENDAR DATE	ROTATION NUMBER	HELIOGRAPHIC			DIAMETER	RA	DEC
		Lo	Bo	Po	(arcmin)	(HH:MM)	(deg)
1/ 1/2012	2118	117.74	−2.95	2.33	32.531	18:43.9	−23.07
1/ 2/2012	2118	104.57	−3.06	1.85	32.531	18:48.3	−22.99
1/ 3/2012	2118	91.40	−3.18	1.36	32.531	18:52.7	−22.90
1/ 4/2012	2118	78.22	−3.30	0.88	32.531	18:57.1	−22.80
1/ 5/2012	2118	65.05	−3.41	0.39	32.531	19: 1.5	−22.70
1/ 6/2012	2118	51.88	−3.52	−0.09	32.531	19: 5.9	−22.59
1/ 7/2012	2118	38.70	−3.63	−0.57	32.530	19:10.2	−22.47
1/ 8/2012	2118	25.53	−3.75	−1.06	32.530	19:14.6	−22.35
1/ 9/2012	2118	12.36	−3.85	−1.54	32.529	19:19.0	−22.21
1/10/2012	2119	359.18	−3.96	−2.02	32.528	19:23.3	−22.08
1/11/2012	2119	346.01	−4.07	−2.49	32.527	19:27.7	−21.93
1/12/2012	2119	332.90	−4.18	−2.97	32.526	19:32.0	−21.77
1/13/2012	2119	319.73	−4.28	−3.45	32.524	19:36.4	−21.61
1/14/2012	2119	306.56	−4.38	−3.92	32.523	19:40.7	−21.45
1/15/2012	2119	293.39	−4.49	−4.39	32.521	19:45.0	−21.27
1/16/2012	2119	280.22	−4.59	−4.86	32.519	19:49.3	−21.09
1/17/2012	2119	267.04	−4.69	−5.32	32.517	19:53.6	−20.90
1/18/2012	2119	253.87	−4.78	−5.79	32.514	19:57.8	−20.70
1/19/2012	2119	240.70	−4.88	−6.25	32.512	20: 2.1	−20.50
1/20/2012	2119	227.53	−4.97	−6.71	32.509	20: 6.3	−20.29
1/21/2012	2119	214.36	−5.07	−7.16	32.506	20:10.6	−20.08
1/22/2012	2119	201.19	−5.16	−7.61	32.503	20:14.8	−19.86
1/23/2012	2119	188.02	−5.25	−8.06	32.500	20:19.0	−19.63
1/24/2012	2119	174.85	−5.33	−8.50	32.497	20:23.2	−19.40
1/25/2012	2119	161.68	−5.42	−8.95	32.494	20:27.4	−19.16
1/26/2012	2119	148.51	−5.51	−9.38	32.490	20:31.6	−18.92
1/27/2012	2119	135.34	−5.59	−9.82	32.486	20:35.7	−18.67
1/28/2012	2119	122.17	−5.67	−10.25	32.483	20:39.9	−18.41
1/29/2012	2119	109.00	−5.75	−10.67	32.478	20:44.0	−18.15
1/30/2012	2119	95.83	−5.83	−11.09	32.474	20:48.1	−17.88
1/31/2012	2119	82.66	−5.90	−11.51	32.470	20:52.3	−17.61

Index

Index

Printed in the United States of America